GERMAN HERO-SAGAS AND FOLK-TALES

Oxford Myths and Legends in paperback

*

African Myths and Legends *Arnott*
Armenian Folk-tales and Fables *Downing*
Chinese Myths and Fantasies *Birch*
English Fables and Fairy Stories *Reeves*
French Legends, Tales and Fairy Stories *Picard*
German Hero-sagas and Folk-tales *Picard*
Hungarian Folk-tales *Biro*
Indian Tales and Legends *Gray*
Japanese Tales and Legends *McAlpine*
Tales of Ancient Persia *Picard*
Russian Tales and Legends *Downing*
Scandinavian Legends and Folk-tales *Jones*
Scottish Folk-tales and Legends *Wilson*
Turkish Folk-tales *Walker*
West Indian Folk-tales *Sherlock*
The Iliad *Picard*
The Odyssey *Picard*
Gods and Men *Bailey, McLeish, Spearman*

German Hero-sagas and Folk-tales

Retold by
BARBARA
LEONIE PICARD

Illustrated by
JOAN KIDDELL-MONROE

OXFORD UNIVERSITY PRESS
OXFORD NEW YORK TORONTO

Oxford University Press, Walton Street, Oxford OX2 6DP

Oxford New York Toronto
Delhi Bombay Calcutta Madras Karachi
Kuala Lumpur Singapore Hong Kong Tokyo
Nairobi Dar es Salaam Cape Town
Melbourne Auckland Madrid

and associated companies in
Berlin Ibadan

Oxford is a trade mark of Oxford University Press

First published 1958
Reprinted 1959, 1965, 1971
First published in paperback 1993

A CIP catalogue record for this book is available from the British Library

ISBN 0 19 274163 2

Printed in Great Britain
on acid-free paper

To my Mother

E.M.P.

who helped

Contents

HERO-SAGAS

FOLK-TALES

HERO-SAGAS

I

Gudrun

LONG, long ago there ruled in the land of the Hegelings, near the mouth of the River Scheldt, a king named Hettel. He had a fair queen named Hilde, a daughter even fairer than her mother, Gudrun, and a young son, Ortwin.

The fame of Gudrun's beauty spread into many lands, and bold warriors came wooing her from far and wide; but on one of them only did she smile—Herwig, the young king of Zealand.

In the land of Normandy ruled Ludwig and his queen, Gerlinde; and when their son, Hartmut, heard of Gudrun's beauty, he said, 'I will go into Hegeling-land and win Hettel's daughter for my bride, for all he is a mighty king and his lands are wider than ours.' And his mother, Queen Gerlinde, who loved him well and wanted the best of all things for him, said, 'Go, and may you prosper, my son. For only the fairest of all maidens is worthy of you.'

So Hartmut went to the land of the Hegelings with his followers, bearing gifts for King Hettel and Queen Hilde, and they received him kindly. But when he told them why he had come to their land, Hettel frowned and said, 'My daughter is not for you.' And Hilde scorned him, saying, 'My father is the great king of Ireland, and your father holds his lands from mine. Shall my daughter wed with one of my father's vassals?' So Hartmut returned to Normandy, shamed and angered that his wooing was disdained.

After a time, King Hettel promised his daughter to Herwig of Zealand, and Gudrun was well pleased. Herwig returned to his own land to prepare for the marriage feasting, but his rejoicing was soon turned to sorrow, for the king of the Moors invaded his lands with a mighty army, burning and plundering. Herwig and his men fought bravely, but they were sorely pressed, and Herwig sent messengers to King Hettel, asking his help. Immediately, Hettel set out for Zealand with his young son Ortwin and with all his finest warriors: Frute from Denmark, Horant the minstrel, and old Wate, the fiercest of them all. With the help of the Hegelings, the king of the Moors was soon defeated, and he and his remaining men were surrounded in a stronghold they had captured.

But in Normandy Hartmut thought continually of how he had been slighted and swore that he would win Gudrun for himself. When she heard how the Hegelings were fighting in Zealand, Queen Gerlinde said to her son, 'Go now with your warriors to the land of the Hegelings and steal away Gudrun while she is unprotected.'

So King Ludwig called together his men for the sake of his son and sailed with him from Normandy in many ships, along the coast to the mouth of the Scheldt. There they landed and went with speed to the fortress of Matelane, where Queen Hilde with Gudrun awaited the return of King Hettel.

Hartmut sent two of his lords to Matelane, and they demanded to speak with Gudrun. She received them graciously and asked them their errand.

'Lady,' they replied, 'we have come wooing for our king's son, Hartmut. He would have you for his wife.'

'Tell him,' said Gudrun, 'that I am betrothed already. I shall be King Herwig's wife, or no man's.' And she had wine poured for them, that they might drink before they rode away.

But the two lords would not drink. 'Hartmut bade us tell you that if you refuse him, you shall see him and all his warriors before your gates, three days from now.'

Gudrun only laughed at their boastful words; while one of the Hegeling warriors who had been left to guard her and the queen called out, 'If you do not care to drink King Hettel's good wine which has been poured for you, be sure that we can pour your blood instead.'

When Hartmut heard the answers which the Hegelings had given, he was angry; and he and his father rode for Matelane at the head of all their men.

When they saw the Normans drawn up before the fortress, the Hegeling warriors said, 'See, Hartmut of Normandy has come a-wooing with his sword. There will be some broken helmets before nightfall.' And they made ready to go forth to meet their enemies. But Queen Hilde said, 'We are too few to resist them. Close the gates and let us do no more than defend ourselves from their attack. In that way we shall survive until the king returns.'

But the Hegelings scorned her wise counsel and would not be ruled by a woman when there was fighting to be done. 'Shall we hide like cowards and so be shamed for ever?' they asked. And they opened the gates of the fortress and rode out to meet the Normans. But even before the last of them had passed through the gates to slay and be slain, the first of the Norman warriors rode in.

Hartmut went to where Gudrun waited fearfully and took her by the hand. 'Too long have you slighted me and scorned my love,' he said. 'It would not be unjust were I to scorn you in return and hang you, along with everyone in Matelane. But I will be merciful and you shall come to Normandy with me and be my queen.' And he led Gudrun, pale and distraught, down to the Norman ships, followed by her maidens, weeping and afraid.

The Norman warriors seized all Hettel's treasure

from Matelane, and they would have burnt the fortress as well, had not Hartmut forbidden it.

Then the Normans and their captives set sail for Normandy, and Hilde sent word to Hettel of what had befallen, bidding him save Gudrun if he could.

When Hettel and Herwig heard the news they wept, but old Wate said, 'We must make peace with the king of the Moors with no delay, so that we may be free to fight the Normans.' And it was done as he counselled, and within a few days Hettel and Herwig and their warriors were sailing for Normandy.

Half-way to Normandy, King Ludwig and Hartmut landed on the Wülpensand, a North Sea island, and encamped on the shore, believing themselves safe from pursuit. But there the ships of the Hegelings came down upon them, and the warriors of Hettel and Herwig poured from the ships on to the beach, and a great battle they had there, on the Wülpensand.

Brave deeds were done by many in that place: by Frute of Denmark and Horant the minstrel, by Herwig and by young Ortwin, and no one could withstand old Wate's mighty strokes.

In the bitter fighting the two kings, Hettel and Ludwig, met and strove together, giving stroke for stroke, until Hettel was slain by Ludwig. The Hegelings mourned his loss with tears, all save old Wate, who cried aloud for vengeance and rushed upon the foe in wrath, slaying all he found until darkness fell and he could no longer see to kill.

That night the Normans sailed away unseen by the

Hegelings, for there was no moon. When day dawned and Wate rose up, eager for more fighting, and found no foes left save the Norman dead, his fury was terrible.

Sadly the Hegelings and Herwig's men sailed back to Hegelingland; but none dared face Queen Hilde in Matelane to tell her that the king was dead and Gudrun still a captive, save only Wate, and he went alone to her and told her. Hilde wept for her husband dead and her daughter lost, and so many Hegeling warriors slain; but Wate said, 'When those who are but lads today, are grown warriors, we will be avenged on Ludwig and on Hartmut.'

'May I live until that day,' said Hilde.

In their ships, the Normans rejoiced when the coast of Normandy came in sight, and King Ludwig called Gudrun to his side and pointed out his lands. 'See, daughter of Hettel, a share of all my wealth shall be yours, if you are willing.'

But Gudrun answered sadly, 'I want none of it, for my father is dead and I am parted from him who is to be my husband.'

'Grieve for that no longer,' said Ludwig. 'Marry my good son Hartmut and forget your sorrow.'

'Let me be,' said Gudrun. 'I would rather die than marry Hartmut, whose father is no more than a vassal of my grandfather.'

King Ludwig was so angered by her words that he caught her by her golden hair and would have flung her into the sea, had not Hartmut prevented him.

'Had any man but my father dared to lay hands upon my bride, I would have slain him for it,' said Hartmut, frowning.

'I would end my days in peace with my son,' said Ludwig. 'Forbid your bride to anger me.'

Queen Gerlinde and her daughter Ortrun came down to the shore to welcome Ludwig and Hartmut, and they rejoiced when they saw how Hartmut had brought Gudrun home with him. Ortrun kissed her kindly, seeing her so sad; but when Gerlinde would have embraced her, Gudrun drew away. 'I have heard,' she said, 'how it was you who counselled your son to steal me from my father's house and bring me as a captive to a strange land. I want no greeting from you.' Yet she had no choice but to go to the king's house with Gerlinde, she and all her maidens.

The Normans feasted long for their victory, while Gudrun sat among them silently and with a heavy heart, and only Ortrun pitied her and understood her grief.

Then Gerlinde said, 'We should soon be feasting for another cause. When does my son marry his bride?'

Gudrun answered quietly, 'Through him I lost my father, and by his hand died many of the Hegelings. I will not marry Hartmut.'

Proud Gerlinde smiled on her and said, 'Tears will not bring them back to life. Laugh and be merry, Gudrun. Marry with my son and I will give you my crown and my rights.' For, proud though Gerlinde was, her love for Hartmut was greater than her pride, and she wished to see him happy.

'I want neither your son for my husband, nor your crown upon my head,' said Gudrun.

Gerlinde frowned, but Hartmut said, 'I will not have her forced. She must marry with me willingly or not at all.'

Gerlinde smoothed the frown from her face and laid her hand on his arm. 'She is like a foolish, wilful child, my son. Leave her in my care, and I will teach her better ways.'

Hartmut turned and looked at Gudrun where she sat amongst her weeping maidens, pale and cold, staring before her, with her hands clasped in her lap. 'She is a stranger in our land, and unhappy. Be gentle with her, good mother.'

And so Hartmut went from his father's house leaving Gudrun in his mother's care, believing that, for his sake, Gerlinde would have patience with her.

But for his sake, that she might see him happy with the bride whom he had chosen, Gerlinde was harsh with Gudrun, striving with hard words and severity to break her pride and make her yield to his demands. 'If you will not take the happiness you have been offered, then you shall have sorrow instead,' she said. She took Gudrun's maidens from her and set them to work, spinning and weaving and embroidering with silk and golden thread; while Gudrun herself she ordered to fetch wood for her fires and carry water for herself and her women; she gave her ragged clothes to wear and little enough to eat and left her no hour to rest, in all the day.

And Gudrun obeyed and did all that she was commanded to, for she thought it better to serve Gerlinde and Ortrun and their Norman maidens at the lowest tasks, than to marry with an enemy. And of all there, only Ortrun was kind to her, when her mother was not by to see it. Yet always even Ortrun would say to her, 'Dear Gudrun, my brother is a good man and a brave warrior. It would be no shame to you to marry him. And I would gladly call you my sister.'

The months passed and Gudrun grew pale and thin, and then Hartmut returned to his father's house. When he saw her in rags he said, 'What ails you,

Gudrun?' and bitterly she answered him, 'I have to work as a servant for your mother. See to what shame and misery you brought me when you stole me from my father's house.'

Hartmut was distressed for her, and angry, and he reproached his mother. 'I bade you treat her gently and I trusted you. And you have made a serving-wench of her.'

'How else would you have me break her stubbornness and pride?' asked Gerlinde. 'I would see her married with you soon.'

'Be harsh with her no longer, my mother. And treat her fittingly, as a king's daughter should be treated.'

'Do not fear,' said Gerlinde. 'I will do my best for you. She will yet be your wife.'

And once again Hartmut trusted his mother and left Gudrun in her care, taking leave of Gudrun with a smile. 'I shall come back to claim you for my bride and you shall wear a crown and be happy at my side.'

'I am betrothed already,' said Gudrun, 'and to a greater king than you will ever be.'

But when he was gone, Gerlinde, in anger, sent for Gudrun. 'You would make strife between my son and me, but it will not avail you. I shall find even lower tasks for you until you yield to our demands.' And she sent Gudrun down to the shore, to kneel all day in the sun or in the bitter cold to do the washing for her household.

Gudrun's maidens wept when they saw her so mistreated, all save two of them, Hergart and Hildeburg.

Hergart was no longer unhappy in her captivity, for King Ludwig's cupbearer loved her and had promised to marry her; while Hildeburg, angry and indignant, stood up boldly among the others and said aloud, in the hearing of Queen Gerlinde, 'Well may you weep, all of you. For what pity or respect may we, her maidens, expect, when our mistress is shown so little.'

'You shall suffer for your words,' cried Gerlinde furiously. 'Go out and help your mistress. When the snow lies on the ground you will wish you had not spoken rashly.'

But Hildeburg went joyfully to Gudrun on the shore, and Gudrun was glad of her company. 'The time will pass less slowly,' she said, 'if you are with me, and we can cheer each other as we work, with talk of our home and with tales.'

So, every morning at sunrise, summer and winter alike, Gudrun and Hildeburg carried the baskets of garments down to the shore and washed all day; and every evening at sunset, they carried them back to King Ludwig's house. And so it went on, until Gudrun had been a captive in Normandy for seven years.

And when seven years were passed, in Hegelingland old Wate went to Queen Hilde and said, 'Those who were no more than boys when King Hettel died are now warriors. It is time we went to Normandy and fetched Gudrun home.' Hilde sent messengers to all who owed service to her and to King Ortwin her son, and they came with their men: Frute of Denmark, Horant the minstrel; while Herwig came from his kingdom of Zealand to help win back his bride. Together they all set sail for Normandy.

On the coast of Normandy they landed and encamped close by a wood in a wild spot where none noticed their coming to warn King Ludwig of it. 'It would be best,' said Frute, 'that we sent secretly to find how Gudrun fares, before we attack our enemies.' Though old Wate was ready to fight at once, with no delaying.

'I will go alone to learn how Gudrun fares,' said young Ortwin. 'She is my sister, so it is my duty and my right.'

'Gudrun is my betrothed wife,' said Herwig, 'so mine is the duty and the right.'

Ortwin smiled. 'Then let us go together,' he said.

'Young fools,' grumbled Wate. 'If Ludwig catches you, he will hang you both.'

But Ortwin and Herwig set off alone, no more than the two of them in a little boat, making their way farther along the coast of Normandy, to learn what they could of Gudrun.

And it so happened that one morning, in the bitter cold of early spring, when snow still lay upon the ground, as Gudrun and Hildeburg washed clothes upon the beach, they looked up and saw a little boat, making for the shore. 'It would be a lasting shame to me that a stranger should find me, the daughter of a king, a ragged washerwoman,' said Gudrun. 'Let us hide, Hildeburg, until they have landed and gone past.'

But as they leapt ashore from their little boat,

Ortwin and Herwig called out to them to come back, promising them no harm. 'We are strangers,' said Herwig. 'Tell us what land this is, and who rules here.'

And Gudrun knew his voice, though she dared not believe he had come at last. 'King Ludwig of Normandy and Hartmut his son rule here,' she said. But Herwig did not know her, for she was in rags and barefoot, with her hair blown across her face by the wind.

'Can you tell us of a company of captive maidens whom King Ludwig brought here, many years ago?' asked Herwig. 'And Gudrun, whom they served, is she alive or dead?'

'If you ever saw her, you will remember her,' said Ortwin. 'I knew her, in the days when I was no more than a boy and she was the fairest of all maidens.'

'I am she,' said Gudrun; and with a cry of joy, Herwig took her in his arms. But Ortwin said, 'Your husband Hartmut uses you ill, good sister, if he gives you rags to wear and lets you wash clothes upon the shore.'

'He is not my husband,' said Gudrun, 'and for that reason his mother Queen Gerlinde treats me as a serving-wench.'

'In spite of Wate's doubts, our fortune could not have been better,' laughed Herwig. 'Let us take Gudrun and this other maiden with us now. The boat will hold the four of us.'

But Ortwin would hear none of it. 'I would rather

die, and I would rather my sister died also, than that I should steal away secretly from any man that which he took from me openly in battle. And besides, what of the other maidens, shall we leave them here in captivity? And shall we be cheated of our vengeance?' He kissed Gudrun. 'It is not that I do not love and honour you, dear sister, but I would not be shamed in other men's eyes. I will come again tomorrow, openly, with all our warriors.'

With tears Gudrun and Hildeburg watched them row away, and when they could no longer see the little boat Hildeburg sighed and dried her tears and went on with the washing. But Gudrun sat on a rock in silence, looking out to sea.

After a time Hildeburg said, 'Dear mistress, it grows late and Gerlinde will be angry with us if the work is not done. Why do you not help me?'

'Never again shall I wash for Gerlinde,' said Gudrun. 'I am too high to be her servant. Today two kings have held me in their arms and kissed me.'

'Gerlinde will have us beaten if the clothes are not washed,' said Hildeburg.

'Tomorrow comes freedom and vengeance. Let them beat me all night if they will. I shall not die of it.' Gudrun laughed. 'Today I am a king's daughter again. Let Gerlinde's clothes share in my freedom.' She rose and took up an armful of garments and, before Hildeburg could prevent her, she had flung them far into the sea, so that they were lost, every one.

At sunset Hildeburg, laden with baskets and washing, trudged up from the shore; and behind her walked Gudrun, empty handed. In the doorway of the king's house they found Gerlinde waiting for them. 'You are late,' she said. 'You have been idling. I shall see to it that you are sorry.'

'You would be wisest not to threaten me,' said Gudrun calmly, 'for I am more highly born than you.'

'You are insolent,' said Gerlinde angrily. 'And where is the other linen you took with you this morning?'

Gudrun shrugged her shoulders. 'I do not know. Nor do I care. I threw it into the sea.'

Gerlinde, enraged, sent for rods and would have beaten Gudrun, but a sudden crafty thought came into Gudrun's mind. 'If you beat me today,' she said, 'I shall see that you pay for it when I sit beside Hartmut as his queen. For I am tired of serving you and I will marry your son. Send to him and bid him come here and marry me tomorrow.'

Gerlinde was so pleased at her words that she had no suspicion of her intent. She sent at once to Hartmut and he rode that very night for his father's house and went to Gudrun with a light heart. But when he would have taken her in his arms she prevented him, saying, 'It would be a shame to you to kiss a bride unkempt and in rags. Tomorrow, when I am apparelled like a queen, you shall embrace me.'

In his joy he said, 'All things shall be as you wish. Tomorrow shall be happiness and feasting, for I have at last won the bride I set my heart upon.' And Gudrun's joy was no less than his, for she had brought him to his father's side, and thus he would not escape the vengeance that would come in the morning.

And early in the morning, at the first light of day, looking from King Ludwig's house, his watchmen saw the Hegelings and the men of Zealand all about them, their spears glinting cold in the dawn.

'Gudrun's kinsmen are come to fetch her home,' said Hartmut bitterly. 'Small joy have I had of her in seven years.'

King Ludwig and Hartmut rode out through the gates at the head of their men, and met the onrush of the Hegelings and Herwig's men; but they went too far from the protection of their gates, and for all their courage and battle-skill, they were overcome in the end.

Old Wate fought grimly, hewing down right and left of him, caring little how many good warriors were slain, so long as the Hegelings had their vengeance; and in the battle Herwig met with King Ludwig and they fought together. For all Ludwig was an older man, he was well matched with Herwig, and more than once, as they struggled, he almost worsted him, while Gudrun watched fearfully from the walls. But at last, with a mighty stroke of his sword, Herwig cut off Ludwig's head, and King Hettel was avenged.

Slowly the attackers gained the advantage, until Hartmut saw that he could not hope to win the day. He called his remaining men about him. 'We must

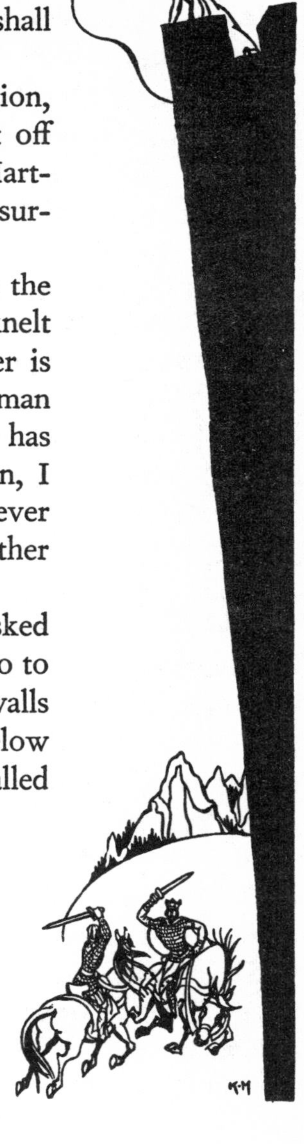

make for the gates and close them after us,' he said, 'and so we shall live to avenge this day.'

But Wate guessed his intention, and with his own warriors cut off the way to the gates, so that Hartmut and his Normans were surrounded.

When Ortrun saw this from the walls she ran to Gudrun and knelt at her feet, weeping. 'My father is slain, and all our good Norman warriors; and now my brother has but little time to live. Gudrun, I have always pitied you and never done you harm; save my brother Hartmut for me.'

'How can I help him?' asked Gudrun. 'I am no warrior to go to his aid.' Yet she went on the walls and looked out, and there below the walls she saw Herwig and called to him to save Hartmut from Wate for Ortrun's sake. Herwig hastened to where Hartmut and Wate fought before the gates, and he called to Wate to spare Hartmut's life. 'For my bride and his sister ask it of you,' he said.

'Shall I care what women counsel?' exclaimed Wate. 'If I have my way, he and all his kin shall die.'

But, in spite of Wate, Hartmut was taken alive, and led captive to the ships, even as the Hegelings surged into the king's house, with Wate at their head, slaying all who stood in their way.

In the queen's bower all the women of Ludwig's house waited together fearfully. Even Gudrun and her own maidens who crowded about her were afraid as they heard the din of battle coming closer, and heard old Wate's voice roaring encouragement to his warriors as they hunted out those foes who sought to hide themselves from death.

Gudrun looked across the room and held out her hand to Ortrun. 'Come, stand beside me and I will save you from the Hegelings.' And Ortrun ran from among the trembling Norman maidens and took Gudrun's hand.

Gerlinde, white-faced with grief and fear, came to kneel before Gudrun. 'Save me,' she begged, 'from Wate and his men.'

'Have you ever shown me kindness, that I should show you any now? Ask nothing of me, Queen Gerlinde, for I hate you with all my heart.' And Gudrun turned away from her, and Gerlinde rose and went back to stand amongst her own women, knowing that there would be no mercy for her.

The shouts and cries came nearer, and then Wate flung open the door and stepped into the room, his reddened sword dripping in his hand. 'Where is Gudrun?' he said.

She stepped forward. 'I am here, alive and safe, good Wate.'

He pointed with his sword. 'Who are those women standing beside you?'

'They are the maidens who were brought with me from our own land. And this is Ortrun, who is my friend.'

'And where is that she-wolf, Gerlinde?' Wate walked the length of the bower seeking her; and Hergart, who had married King Ludwig's cupbearer, edged away from the queen's women and ran swiftly to Gudrun. 'Have pity on me, or I shall be slain.'

Angrily Gudrun said, 'You have yourself to thank for that. Had you shared in our misery, you could share in our joy today. Why should I care what becomes of you?' She turned her head away; but after a moment she said, 'Stand closer to me, and perhaps you will be overlooked.'

Wate returned to Gudrun. 'Where is Queen Gerlinde who made you wash clothes upon the shore? Where are Gerlinde and her kinswomen?'

Gudrun looked straight before her, neither to the right nor to the left. 'I do not see them here,' she said.

Wate's wrath was terrible. 'I will slay all the Norman women if I do not find their queen.' And one of Gerlinde's women, to save herself, made him a sign, so that he knew which one of them was the queen, and he went in amongst them and fetched her out. 'No more shall the sister of my king wash clothes for you,' he said. He dragged her from the room by her hair, and outside the door he struck off her head. Returning, he said, 'I have heard that one of the Hegeling maidens wedded with Ludwig's cupbearer. Where is she?'

All Gudrun's maidens cried for mercy for her and would not tell him which she was; but Wate knew Hergart by her terror and by her richer garments and her jewels, and though she hid herself behind Gudrun, he found her and slew her where she stood. 'Now is our vengeance completed,' said Wate, and sheathed his sword.

Then Ortwin and Frute and Horant the minstrel came to Gudrun and she welcomed them gladly; and when Herwig came, who had fought so bravely for her that day, her joy was perfect.

For many days the Hegelings and the men of Zealand feasted in Normandy, and then, with their booty and their prisoners, they returned to Queen Hilde, who greeted them with joy; and in Matelane was Gudrun at last married to King Herwig. And that there might be peace in that happy time between the Hegelings and the Normans, Ortwin was betrothed to Ortrun, to Gudrun's great content, and Hartmut was set free and he returned to his own land to rule as king.

II

Dietrich of Bern

IN the early days, south of Germany, a king named Dietmar ruled over the Amelungs in the little kingdom of Bern. His elder brother, Ermenrich, was lord of a mighty empire, but Dietmar was content enough with the love and the loyalty of his Amelungs and he acknowledged no man as his master.

His great pride was his son Dietrich, a bold and handsome child, who soon grew tall and strong beyond his years. When he was five years old, his father gave him into the keeping of Hildebrand, the greatest warrior in those parts, that he might train the boy in battlecraft and feats of arms. And so skilled was his teaching and so eager was the boy to learn, that by the time he was twelve, Dietrich was as strong and able as any warrior of twice his years. From the first day that he saw him, Dietrich loved and honoured Hildebrand, and they were firm friends for all their lives.

When Dietrich had grown to be a youth, there came to the king's house one day word of two giants, Grim and his sister, Hilde, who were ravaging the land and killing all who tried to withstand them. King Dietmar immediately set out with his warriors to slay them; but though he searched his kingdom from end to end, the giants had learnt of his coming and had hidden themselves too well, high up in the mountains, and he could not find them. Dietmar returned home, discouraged and angry, and, immediately, the giants came out of hiding and began to plunder farms and villages once more.

Dietrich said to Hildebrand, 'Let us go alone, just you and I, and search out these monsters in their lair. Who knows, we might have the luck that was denied my father.'

So Hildebrand and young Dietrich set off; but for a long time their search seemed hopeless. And then one day, in the mountains, Dietrich chanced to catch one of the dwarf folk who lived beneath the earth.

'Keep a fast hold of him,' said Hildebrand. 'He may well be a friend of the giants, and can tell us where they are.'

But the dwarf swore that he was no friend to the giants, who had done him and his kind much wrong. 'I am Elbegast, the lord of the mountain dwarfs,' he said. 'If you are foes of Grim and Hilde, then you are friends of mine.' And he promised to lead them to where the giants might be found. 'But,' he said, 'you will never slay them without the help of a weapon forged by the dwarfs. For the dwarfs are the finest swordsmiths of all.'

So he gave to Dietrich the sword Nagelring, which had no equal in the world. Then, early one morning, he took Dietrich and Hildebrand by a hidden path to where they could see the giants' footprints, huge tracks upon the dewy grass, leading to the hollow mountain where they lived. 'Good luck be with you both,' said Elbegast. 'And may Nagelring prove trusty.'

Dietrich and Hildebrand followed the footprints to the great cave where the giants were hidden; but, as they reached the mouth of the cave, Grim heard them and rushed forth like a mountain tempest, brandishing above his hideous head a burning log snatched up from his fire. With this huge cudgel he struck at Dietrich again and again; and had Dietrich not been quick and light upon his feet, he would have been dead in a very little time. But not a chance did Dietrich have to strike a single blow with Nagelring in return, for he needed all his strength and wits to avoid the giant's blows and the sparks which flew from the smouldering log.

Hildebrand would have gone to his aid, had not Hilde, hearing the sounds of the combat, come from an inner cave, and before he could strike at her with his sword, she had caught him up in her two arms and crushed him to her as though she would break all his bones.

Hildebrand struggled in vain in her grip. Closer

and closer she clasped him until he thought that death was surely near, and with his little remaining breath he gasped out, 'Dietrich, help me, or I am dead.'

Dietrich heard, gave a swift glance round and saw how it was with his friend and, made desperately bold by his fears for Hildebrand, he leapt right over the flaming club as it was swung at him again and brought Nagelring down with all his strength upon the giant's head, splitting his thick skull. Then he turned to Hilde, and before she could fling down Hildebrand to defend herself, he had slain her too.

When he was a little recovered and could speak once more, Hildebrand smiled and said, 'I taught you all I know of skill at arms, yet it seems you have surpassed me and could now teach me much.'

All the people of Bern rejoiced when they heard that the giants were slain and would trouble them no longer, and King Dietmar was more than ever proud of his son, who had acquitted himself so well on his first adventure.

But soon after, to the great sorrow of the Amelungs, King Dietmar died, and so young Dietrich became king in Bern.

A few years passed and then one day Dietrich learnt of a third giant, Sigenot, more terrible by far than the others, who had entered Bern and was slaying cattle and people alike in vengeance for the death of his kinsfolk Grim and Hilde.

Dietrich called for Nagelring to be brought to him. 'I will go to the mountains and slay this Sigenot,' he said.

Everyone who heard him, save Hildebrand, cried out, 'You must not go. He is too terrible. Not even an army could withstand him.'

But Hildebrand said, 'You shall not go alone, for I will go with you, my king.'

'No,' said Dietrich, 'for he is only one. One warrior against another, that is the law of fair combat. Two against one is the cowards' way. You taught me that yourself.'

'Then go alone,' said Hildebrand, 'and may good fortune go with you. If you have not returned when eight days are passed, I will go after you to free you if you are a prisoner, or to avenge you if you are dead, or to die myself at the giant's hands.' They embraced, and Dietrich went.

He tracked Sigenot to a cave in the mountains, and there he came face to face with him; and Sigenot was even taller and broader than Grim had been. With a roar he took up his great club and strode to where Dietrich stood. 'You are Dietrich, for no one else would be bold enough to come to me alone,' he said. 'Now shall my kinsfolk be avenged.' And he swung his club above his head.

As he had done with Grim's flaming brand, Dietrich darted here and there to avoid the mighty blows of the club; at the same time giving stroke after stroke with Nagelring, until it seemed as though he might, in spite of the giant's huge bulk, be the victor in their combat. But then, dodging quickly to one side, he came beneath a tree, and as he raised Nagelring high above his head with both his hands to strike at Sigenot once more, the blade caught in the overhanging branches and, before he could free it, Sigenot's club had crashed down upon his helmet and he fell senseless to the ground.

With a great shout of triumph, Sigenot snatched up Dietrich in his arms, flung him over his shoulder, and strode off to his cave; and there, in the darkest corner of the cave, he cast him down into a pit of serpents.

When the eight days were passed and Dietrich had not returned, Hildebrand armed himself, took sword and shield, and rode for the mountains. There he spied the giant's tracks, leading to his cave, and there he found Dietrich's horse wandering alone; and there, too, he saw Nagelring caught fast in the branches of the tree. 'So Dietrich is dead,' he thought. 'Well, I can but avenge him or die too.' And he took a firm hold of his sword and went forward to the cave.

Sigenot saw him approaching and came out, swinging his great club. 'You are Hildebrand,' he shouted.

'First I catch the young one and now I catch the old one. Truly, Grim and Hilde will be well avenged.' And eagerly he rushed upon Hildebrand.

The combat lasted many hours. In his rage, Sigenot tore up young trees to serve as weapons, and heaved up rocks and stones to hurl at Hildebrand; but with skill Hildebrand evaded all his blows, until he was too spent and the giant's mighty strength proved too much for even his endurance and, like Dietrich, he was felled by Sigenot's club. Sigenot slung him over his shoulder, took his sword as a victor's prize, and strode back to the cave, shouting his triumph till the mountain echoed.

He flung Hildebrand to one side of the cave and the sword to the other and went off to find a chain that he might bind his senseless captive. But the force of the fall brought Hildebrand back to himself and he sat up, dazed and battered, and looked about him. The cave was wide and light enough, though in its farther corners no daylight reached, and Hildebrand saw his sword lying where Sigenot had flung it. He staggered to his feet and took it up and hid himself behind a pillar of rock until the giant should return.

When Sigenot came back, rattling and clanking the chains he had fetched, he gave a roar of anger at seeing Hildebrand gone from where he had left him lying. Then he saw him hidden in the shadow behind the rock and rushed at him, and their combat was begun anew.

But Hildebrand was very weary, and step by step he had to give way to the furious giant, and slowly he was forced deeper and deeper into the cave where the light was bad and he could hardly see to avoid the giant's blows. He felt that his strength was going fast, and he knew that he could not fight much longer. 'Now Dietrich will never be avenged,' he thought, 'and in a very little while I, too, shall be dead.' And Sigenot, seeing him weaken, laughed till the cave rang. 'In a moment now, good Grim and Hilde will be avenged,' he cried.

But Dietrich, who had been listening to the sounds of fighting from the cave above him, heard the shout of triumph and called out, 'Is it you, Hildebrand, come as you promised? I have been waiting for you.'

At the joyful knowledge that Dietrich was not dead, Hildebrand felt himself the equal of ten giants; his strength returned for one last mighty effort and a moment later Sigenot was lying vanquished on the floor of the cave, and Hildebrand struck off his head.

'Where are you, Dietrich?' he called into the shadows.

'I am here, in the pit of serpents at the very end of the cave. I have killed many of them, and eaten some for food; but there are still scores left alive, so help me out with all the speed you can.'

With ropes lowered into the pit, Hildebrand drew him out, and as the friends embraced, Dietrich laughed and said, 'After all, you are still my master and can teach me much in the matter of fighting skill. For you overcame Sigenot, but I was overcome by him.'

For many years after that, Dietrich ruled well and

wisely, and he gathered about him a little band of skilled warriors from all corners of the world, who had come to him in Bern, drawn by his fame. Yet always his most loved comrade was Hildebrand.

But Dietrich was not destined to rule in peace for all his days. There came a time when his uncle, the Emperor Ermenrich, urged by evil counsellors, cast greedy eyes upon his nephew's little kingdom, and he sent messengers to Dietrich demanding tribute from him and his lords. Now, Dietrich, like his father Dietmar before him, had never paid tribute to any man, and he and his lords were indignant.

'Tell your master,' the Amelungs said to the messengers, 'that we pay our tribute to our rightful lord, King Dietrich, and to no other.'

'And tell my uncle the emperor,' said Dietrich, 'that if he wants his tribute, he must come and fetch it for himself; and we will pay him with our spearheads and the sharp edges of our swords.'

The angry emperor gathered together a great army and marched against Bern. But Dietrich did not wait to be attacked. At the head of his Amelungs he rode out to meet him; and so unprepared for his coming was the emperor that he was taken by surprise; and Dietrich, falling upon his camp before sunrise one morning, won a victory and checked his advance. Though it was no great victory, it cheered the Amelungs and gave them courage for the dark days they knew must lie ahead.

From the emperor's camp Dietrich took much

booty, which he sent back to Bern in the charge of Hildebrand and five of his most trusted warriors: old Amelolt, Sigeband, Helmschrott, Lindolt and Dietlieb of Styria; and with them was Hildebrand's young nephew Wolfhart. But on the way to Bern they were ambushed and taken captive, and only Dietlieb escaped to bear the news to Dietrich.

The emperor was planning to attack again; every day more warriors joined him, sure of his eventual victory; and many, even, of Dietrich's own men deserted him, believing that Bern was lost. But bitterest of all to Dietrich was the knowledge that his trusted friends, and among them Hildebrand, were in the emperor's hands.

Dietrich still had certain of the emperor's lords whom he had taken captive when he had attacked his uncle's camp, and he sent to Ermenrich with an offer to exchange these men for Hildebrand and his five comrades.

But the emperor sent back scornfully, 'Do as you will with your captives, I care nothing for them. For my part, I intend to hang your warriors unless I have your word that you will give me Bern, and that you and those who still wish to call you king will go from the land with you, on foot and leaving all they possess.'

In his first anger at Ermenrich's reply, Dietrich thought, 'I have no hope of victory in battle against his might. Yet he shall at least see how the Amelungs can die. And they would rather die than deign to live dishonoured on his terms.' But then he thought how, if he fought and died gloriously with the few men who remained to him, he was condemning to a shameful death, unfitting to any warrior, his good friends:

Sigeband, Helmschrott, old Amelolt, Lindolt, rash young Wolfhart, and Hildebrand—above all, Hildebrand. And he knew that he could not do it.

He sent to Ermenrich, agreeing to his demands, and Hildebrand and the others were freed. Together, on foot, leading their horses and taking no possessions with them, Dietrich and Hildebrand and their friends left Bern; and with them went those warriors who chose exile with their king rather than service under the emperor: and out of all the Amelungs, there were only three and forty of them.

They wandered northwards to Bechlaren, beside the Danube, where Rüdiger held lands from King Etzel of the Huns. Rüdiger had been a comrade of Dietrich in former days, and he welcomed him kindly and gave him arms and gifts, for he was a good and loyal friend. He gave Dietrich hope, also, for he told him to go to King Etzel and ask his aid. 'He is a mighty king and has many warriors to serve him from all the lands of the world. And though he is a heathen, he is no foe to Christian men. He may well give you help in your fight against the emperor. If not today, then at some later time, when he is in a mood to do so.'

So Dietrich and his few faithful Amelungs went to Hunland to King Etzel's court, and he received them kindly, for he had heard—as who had not—of the prowess and the sad fate of Dietrich. And in time he gave help to Dietrich, and men to fight for him; and in time, too, Ermenrich died and Dietrich returned to Bern and was welcomed by his people. But of the three and forty Amelungs who had gone into Hunland with him, only one—old Hildebrand—returned home with him, for the others had all by mishap been slain at Etzel's court.

Dietrich became not only king of Bern once more, but his uncle, Ermenrich, having had no other heir, he succeeded to the emperor's lands, and ruled them to the end of his days, honoured by all.

III

Walther of Aquitaine

WHEN he was young and eager for conquest, Etzel, the great king of the Huns, rode westwards with his men in search of plunder. From the small kingdoms of the west he took great treasure and much gold as the price of his friendship; and from those kings who asked for time to collect the gold, he took hostages, so that they should not forget to pay him all that they had promised.

From the king of the Franks he took his little daughter, Hildegund; while from the king of Aquitaine he took his young son, Walther. But Gibich, the king of Burgundy, had as yet no children, so he gave to Etzel the lad Hagen of Troneg, his kinsman and the most highly born of all his vassals.

Etzel and his men rode back into Hunland, their wagons and their packhorses laden with spoils, and satisfaction in their hearts at the thought of yet more to come. And with the wild, heathen Huns went Walther and Hildegund and Hagen.

The three children grew up in Etzel's court, where the two boys learnt to ride as only the Huns could ride; while Hildegund span and wove with Queen Helka's maidens. Walther and Hagen became firm friends and were together always; and, whenever she could, Hildegund would steal away from amongst the dark-haired, flat-faced Hunnish maidens and talk with Walther and Hagen, who reminded her of her own people and her home.

After a time, there came a day when King Gibich had sent into Hunland all the gold that he had promised, and Hagen was free to go home. King Etzel was sorry to lose him, for he always welcomed to his court brave warriors from other lands to serve him in his wars, and Hagen had grown into a strong and mighty warrior. But Hagen would not stay in Hunland, for all Etzel's wishes, and so Etzel gave him gifts and sent him on his way with all honour; and he returned home once more to Burgundy.

Walther and Hildegund, left in Hunland, sought each other's company more and more, and in time the two of them, so different from the black-eyed, sallow-skinned Huns they lived among, grew to love each other, and promised one another secretly that if ever they went safely home again they would persuade their fathers to allow their marriage.

But King Etzel, when he was merry with wine or in

a jovial mood, would laugh and say to Walther, 'The little boy I brought from Aquitaine has grown into a fine strong man and handsome besides. It is time you chose yourself a bride, young Walther. Is there one of our Hunnish maids who takes your fancy? You have only to ask, and she is yours. For though you are not of our people, we like you well.'

But Walther would always answer, 'I thank you, King Etzel, but there is time enough for that.' Until he had answered thus so often that the Hunnish lords and warriors began to look askance at him, thinking he slighted them.

And one day Queen Helka said to Etzel, 'Young Hildegund, with her yellow hair, is fairer by far than any of our Hunnish maids. She would be a prize for any man. Have you no kinsman or brave warrior whom you would wish to reward?' So Etzel chose a husband for Hildegund and would have married her to him right away but that she pleaded for a short delay.

'What shall we do?' she said to Walther. 'For now, though you may go home when the time comes, I shall never leave Hunland again, if I am the wife of a Hunnish lord. And how can I ever be your wife, if I have to marry a Hun?' And she began to weep.

'There is only one thing for us to do,' said Walther. 'We must both leave Hunland, and soon. Just the two of us together. Will you be afraid?'

Hildegund dried her tears. 'I will be afraid of nothing, so long as you are with me.' She thought for a moment and then said eagerly, 'I know where the keys are kept to Etzel's treasure house. When we leave, let us take with us some of the gold our fathers have sent into Hunland, for it should have been ours by right.'

One night soon after, while Etzel and his men were feasting and making merry, Walther and Hildegund, on one horse, rode away from King Etzel's house in the darkness, and with them they took two large bags of gold.

The ride out of Hunland was hard, for they had to ride fast, and they had little enough to eat and only the bare ground to sleep on, for they dared trust no one to ask food or shelter of him. But once out of Etzel's kingdom and into Christian lands again, they went slowly and rested often and accepted gratefully all shelter and hospitality which they were offered on their way, telling who they were and whence they had come to any who asked them. Thus it was that word of their escape from Hunland reached Worms in Burgundy while they were yet far from Aquitaine.

Hagen of Troneg smiled when he heard, and was glad, remembering his old comrade Walther; but he said nothing, for he was a man who spoke little, and then to the point.

But young Gunther, who was king in Burgundy, King Gibich being dead, said sullenly, 'My father Gibich paid as much gold to Etzel as Aquitaine and the Frankish king. A third of that gold which Walther carries should be mine by right.'

Hagen looked at him coldly. 'That should it not, as you know full well.'

But Gunther was young, indeed, little more than a boy, and he was wilful and used to having his own way. 'I shall take all the gold from Walther, if I wish,' he said. 'A third, at the least, shall be mine.' He sent for twelve of his best warriors and bade them make ready to go with him to find Walther. 'And you shall come with us,' he said to Hagen. And Hagen shrugged his shoulders and said nothing, but went with them, for Gunther was his king.

In a narrow mountain pass, Walther and Hildegund met with Gunther and his warriors. Gunther called out to Walther to give up the gold he carried. 'For I am Gunther of Burgundy, and I demand it,' he said.

'The gold belongs to me, King Gunther, and to no man else,' said Walther. 'Yet will I give you as much as will lie on a shield, if you will let me pass in peace.'

'Give me all the gold, and you may pass,' said Gunther.

'Come and take it, if you dare,' cried Walther. He dismounted, telling Hildegund to lead the horse away to a place of safety; then, with his sword in his hands, he waited.

Such was the narrowness of the path, that no more than one of the Burgundians could attack Walther at one time. As the first of the warriors fell upon him, Hildegund turned away in fear; but Gunther and his other eleven warriors looked on eagerly, calling out to encourage their comrade. Only Hagen was silent, standing a short way off, leaning on his sword, watching everything; for he alone of the Burgundians had no intention of fighting, since Walther was his former friend.

The combat was fierce, but soon over, and the first of the Burgundian warriors lay dead at Walther's feet.

'He fights well, this Walther,' said Gunther, and sent the second of his men up the rocky path.

One after the other, Walther slew the Burgundian warriors, while Gunther ground his teeth in anger and Hagen still leant upon his sword, watching, with a grim little smile. He regretted the loss of so many good warriors, and in such a cause, but he was not sorry to see Walther acquit himself so well.

At last there came a time when the twelfth of Gunther's men lay dead beside the path, and Walther, looking down, saw only a young lad and his one-time comrade Hagen and, spent and bleeding from many wounds, he prayed that this might be the end and he might now be allowed to pass on his way in peace.

But Gunther, young though he was, was no coward, and with a cry of rage, he unslung the sword from about him and ran up the path against Walther. And Hagen straightened up a little, his hands tightening about the hilt of his own sword, and his eyes narrowed.

Walther was very weary, so his first blow fell wide, missing Gunther, and his sword rang against the ground, striking sparks from the rock. But before Gunther could take the advantage, Walther slashed upwards with his sword and cut deeply into Gunther's leg, so that he slipped and fell and could not rise again. 'Hagen!' he cried. 'Hagen, come to me!'

Walther placed one foot upon Gunther's body and took a fresh grip on his sword. 'It is a pity, King Gunther,' he said, 'that you will die so young, and all for two bags of gold.' He lifted up his sword slowly, for he had little strength left.

But Hagen had come running, before ever Gunther had called to him, and even as Walther would have brought down his sword upon Gunther's unprotected head, Hagen's sword flashed in the sunlight, all but severing one of Walther's arms from his body.

An arm hanging uselessly beside him, his blood streaming to the ground, half blinded by pain, Walther held tightly to his sword with his one hand and, in a last effort, raised the heavy weapon and struck wildly out at Hagen's head. The point of the sword caught Hagen high above the cheek bone, putting out an eye.

So ended a battle for gold that had cost the lives of twelve good warriors, and might have cost three lives more. And there, on the narrow pathway, while Hildegund tended them and bound up their wounds, Walther made peace with Gunther and Hagen and they swore friendship together; and that friendship they never broke.

IV

Siegfried

I

The Slaying of Siegfried

IN the Netherlands there once lived a king named Siegmund and his queen, Sieglinde, in the town of Xanten, beside the Rhine. They had a son named Siegfried who was early famed for his beauty and his courage and his skill at arms. Many adventures Siegfried had, both in his own home and in other lands, and when he was yet little more than a youth he was already respected as a great warrior, and justly, for he had won for himself a hoard of gold, greater than any in the world; he had taken from a dwarf in the northern mountains the Tarnkappe, a cloak which hid its wearer from the sight of all men; and had forged for himself a sword, Balmung, with a green jasper in its golden hilt, and with Balmung he had slain a dragon. It was said of him, moreover, that he had bathed in the blood of this dragon and so rendered himself proof against all weapons, save in one spot between his shoulder-blades where a leaf had fallen on his back while he fought with the monster.

One day Siegfried heard of the lovely maiden Kriemhild of Burgundy, who dwelt in Worms by the Rhine, and one spring morning, with a few trusted warriors, he set out to see her for himself. Gaily and happily he travelled up the Rhine until he came to Worms in Burgundy, where King Gunther ruled with his brothers Gernot and Giselher. There, too, lived Queen Ute, their mother, and Kriemhild, their sister.

Gunther and his brothers welcomed Siegfried kindly, for his fame had travelled to Worms long before he himself came there; and for his part, Siegfried found Kriemhild even lovelier than he had hoped, and he determined that she should be his bride. And thus for a year he remained in Worms with his new friends, and the days passed pleasantly enough.

Now, there ruled at that time in Eisenland, northwards beyond the sea, a fair young queen named Brunhilde, who was as skilled in feats of arms as any warrior; and though many men came to woo her for her beauty, she swore that she would marry none save the man who could cast a spear with more force than she, and who could throw a heavy stone and leap after it for a longer distance. In these three things a suitor must surpass her or lose his life. Gunther had in secret long wished to sail to Eisenland and try his luck at winning Brunhilde, yet he doubted his skill and feared to lose his life. But one day he confessed his secret wish to Siegfried and asked for his counsel.

'Go to Eisenland,' said Siegfried. 'Go to Eisenland and I will go with you. I have my Tarnkappe. Wearing it, no man can see me. In my Tarnkappe I shall stand beside you and help you, unseen. I have no doubt that the two of us together will prove stronger than Brunhilde.' He laughed and cut short Gunther's thanks. 'But none of this will I do for you, my friend, if I may not have your sister for my wife.'

'I could wish for no better lord for her,' said Gunther.

And so it was settled between them, and Gunther and Siegfried set sail for Eisenland, taking with them only one-eyed Hagen of Troneg, and Dankwart, Hagen's younger brother. Down the Rhine they went, to the sea, and on the twelfth day of their voyage, they reached Eisenland.

'In Brunhilde's court,' said Siegfried to Gunther, 'I shall say I am your vassal, even as Hagen and Dankwart are. So shall there be more glory to you in our enterprise, and I shall be the less suspected when I give you my aid.'

Brunhilde received the four strangers proudly, and the looks she cast on them were haughty and unwelcoming. And they saw how she was as tall as any man, and beautiful.

Siegfried spoke for Gunther. 'My lord is the king of Burgundy,' he said, 'and we, his men, are come to

Eisenland with him, for he would win you as his bride.'

'Does he know the conditions?' asked Brunhilde coldly. 'He must cast my spear with more skill than I; he must throw the great stone and leap farther than I. If he fails, he loses his head, and you three die with him.'

'This is folly,' grumbled Hagen to his brother. 'The king will be lucky if he wins safely back to Worms, let alone that he wins a bride to take with him.'

'I know the conditions,' said Gunther, 'and I am confident of my success.'

Brunhilde smiled scornfully. 'Come, then, and try your skill.' She armed herself in a coat of mail and took up her shield of gold and iron.

Siegfried bowed before Gunther and said aloud for all to hear, 'Give me leave to return to the ship, my king.' He went; and on the ship he put on his Tarnkappe and came swiftly back again, unseen of any.

Brunhilde's men brought her her spear, long and sharp and heavy, and fetched a huge millstone and laid it by her; and she waited disdainfully while Gunther armed himself.

When he saw how she had no fear of his skill, Gunther was afraid and wished that he had never come to Eisenland. 'Willingly would I leave her for another man to win,' he thought. But Siegfried came up to him and touched his hand and whispered, 'I am here beside you, Gunther. I will hold the shield before you. Never fear.'

Brunhilde took up her spear and hurled it against Gunther with so much force that it passed right through his shield and would have stretched him on the ground had not Siegfried been there to help him.

'Take up the spear,' whispered Siegfried, 'and aim at Brunhilde. But cast the spear shaft foremost, for we do not wish to kill her.'

Gunther took up the spear and held it poised and Siegfried guided his throw. And together they cast the spear so mightily that Brunhilde fell to the ground. A moment later, and she was on her feet again. 'That was a fine blow, King Gunther,' she said. For though she was angry at her defeat, she admired Gunther for his strength: not knowing that the strength was Siegfried's.

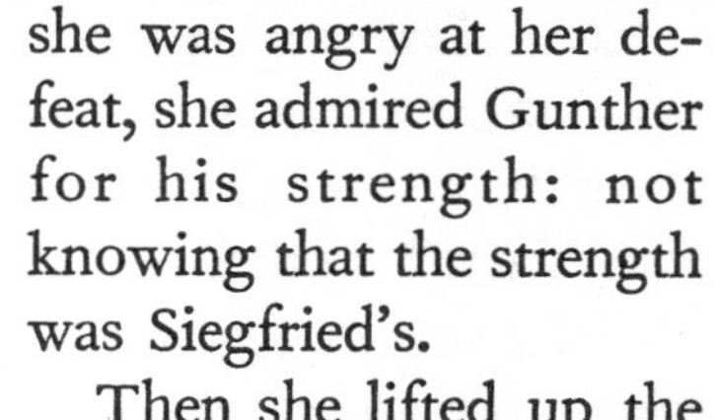

Then she lifted up the great stone and flung it far from her and leapt after it. Four and twenty paces she flung it, and leapt even farther. 'Beat that, if you can, King Gunther,' she said.

Gunther took up the stone with his own hands, but Siegfried threw it, and it fell well beyond the mark set by Brunhilde.

'Now leap, my friend,' said Siegfried, and he and Gunther leapt together; and with his great strength, Siegfried carried Gunther on with him, farther than Gunther could have leapt alone.

Brunhilde saw that she was beaten by the king of Burgundy and she said bitterly, 'You have won your wife, King Gunther.' He went to her and took her hand and kissed her gladly.

But Siegfried hurried to the ship, took off the Tarnkappe, and came back again and made as though he did not know what had befallen. 'When are the tests of strength to begin?' he asked.

'They are over and done with,' said Brunhilde, 'and your king has won. Where were you, that you saw them not? A fine vassal you are, not to stand near your lord at such a time.'

That day, in Eisenland, they feasted the betrothal of Brunhilde and Gunther; and the only one who mourned, though she hid it well, was Brunhilde, for she was angry that she had been won at last. Yet she gained a measure of comfort from the belief that he was so strong and mighty a man who had won her.

Siegfried and the Burgundians returned to Worms, taking Brunhilde with them, and there in Worms, on the day that Gunther married Brunhilde, Siegfried and Kriemhild were wedded also. And Kriemhild rejoiced, for she loved Siegfried with all her heart, and her love lasted until the day of her death. But proud Brunhilde said to Gunther, 'You do ill, my lord, to give your fair sister to a vassal. The sister of the king of Burgundy deserves a better match.'

Gunther laughed. 'Siegfried is my good friend,' he said. 'He deserves the best I have to offer.'

When the festivities for the marriage were over, Siegfried took his bride home with him to Xanten, happy that he had so lovely and so loving a wife. But because he was himself by nature frank and open and the most loyal of friends, and because he understood nothing of women's spite, he saw no harm in telling Kriemhild, since she was Gunther's sister, of how he had helped Gunther win Brunhilde; and never guessed that by this folly he would lose his life.

When some years had passed, Brunhilde said to Gunther, 'Bid your vassal Siegfried come to Worms and bring with him Kriemhild your sister, for I liked her well and would see her again.'

Gunther smiled to himself, that Brunhilde still believed Siegfried his vassal, but he sent to the Netherlands, and bade Siegfried visit him. And so, for the second time, Siegfried journeyed to Worms, glad that he would once again see his friends, Gunther and Gernot and young Giselher.

At Worms there was feasting and minstrelsy and combats between the warriors to entertain the guests. And Siegfried and Gunther joined in the combats, and always Siegfried was the victor.

Watching from where she sat beside Brunhilde, Kriemhild exclaimed, 'I have the best husband in the world. See, how he excels all the other warriors.'

'All, save your brother Gunther,' replied Brunhilde.

Kriemhild laughed. 'Why, my Siegfried easily surpasses Gunther. Truly, I am the luckiest of wives.'

'Gunther,' said Brunhilde, 'is the finest warrior of them all. Did he not win me by his great strength and skill? And even were your Siegfried's strength as great as his—which I do not grant—still would my Gunther be the nobler man, for he is a king and a leader of many men; and Siegfried is no more than his vassal.'

Stung by Brunhilde's scorn and forgetting all her promises of silence in her desire to hurt and humble her, Kriemhild retorted, 'Siegfried is no man's vassal. But if he were, then should you be the wife of a vassal, for it was Siegfried who won you for a bride, not Gunther. Gunther was not strong enough nor skilled enough, so Siegfried, in his Tarnkappe, won you for him.'

Brunhilde stared at her. 'Is this true, which you have told me?' she asked at last.

Kriemhild, rejoicing in her triumph, answered, 'I swear that it is true. I had it from Siegfried himself.'

Brunhilde rose, tall and straight, her face white with anger. 'Siegfried shall pay for this,' she said. In her great rage, that she had been mocked by Kriemhild and cheated by Gunther and Siegfried, she went at once to Gunther. 'This Siegfried, whom you call your friend,' she said, 'he is no friend to you. He boasts openly how it was he who won me as a bride in Eisenland, and says that I should be his wife and not yours. He is treacherous and dangerous, this pretended friend of yours.

You must silence him, my husband, and put an end to his lying for ever.'

At first Gunther laughed at her, but she spoke against Siegfried until at last he began to believe her and he sent for Siegfried, who came eagerly at the summons of his friend. Gunther looked once at Siegfried and then looked away again. 'The queen tells me,' he said, 'that you have boasted openly that it was you, not I, who won her as a bride, and that she should be your wife instead of mine. Can you deny this?'

Siegfried was astonished. 'It is untrue,' he said. 'I have never boasted such a shameful thing to any man.'

'Swear it before all my warriors,' said Gunther, 'and I will believe you.'

So, before all the lords of Burgundy, Siegfried swore his words were true, and Gunther believed him for the time, and smiled on him; and so the matter seemed ended, and Siegfried thought no more of it than just a foolish women's quarrel, and would not have dreamt that the friendship of men could be broken by women's prattling.

But Gunther could not make up his mind one way or the other. When he looked at Siegfried he considered him his loved and trusted comrade; but when he listened to Brunhilde he believed him treacherous and wished him dead. At last he called his lords and his brothers together secretly and asked their counsel.

Gernot and young Giselher were loud in their scorn of his doubts. 'Siegfried is the best friend we ever had,' they said.

But others of the lords spoke against Siegfried through jealousy; while Hagen, grim and dark, said, 'Guilty or guiltless, it is all one to me; but if he disturbs my king's peace of mind, then he were better dead.'

'He would perhaps be better dead,' said Gunther, hesitantly.

'It would be most shameful murder, to slay a guest,' exclaimed Giselher.

'He is my friend,' said Gunther, 'I cannot kill him.'

'You will not need to kill him, my king,' said Hagen. 'Mine shall be the hand to do the deed, and mine alone shall be the guilt. You will have no blame in the matter.'

And at length Gunther gave way to Hagen's arguments and agreed to Siegfried's death; but Gernot and Giselher would have no part in it. Yet through loyalty to their king and brother, they never warned their friend.

Hagen caused a lying rumour to be spread through all the court, of how the king of the Saxons was coming with his warriors against Burgundy. As soon as the rumour reached Siegfried, he went to Gunther. 'I and all those warriors who came with me from Xanten to be your guests, are at your command,' he said. He spoke joyfully, for he was eager to help his friends. And, immediately, he made ready for battle, bidding Kriemhild see that all he needed was prepared.

But Hagen went to Kriemhild and asked her, 'Sister

of my king, are you not grieved that your husband will soon ride out to battle?'

'Why should I be grieved?' said Kriemhild. 'He is the finest warrior of them all, and, besides, he has bathed in the blood of the dragon he slew, and so is proof against all weapons. He will come to no harm in the battle.'

'You are indeed fortunate in your husband, sister of my king, that you always look to see him come home to you unhurt. Have you no fears at all for him?'

Kriemhild hesitated and then said slowly, 'There is one spot where a linden leaf lay upon his skin, and there the blood of the dragon did not touch him. Sometimes I fear when I think of that.'

'Tell me the spot,' said Hagen. 'Make a mark upon his outer garment that I may know where it is, and I will keep close to Siegfried in the battle and my shield shall ever be ready to protect him where he may be hurt.'

And though Hagen was grim and hard and feared by many, Kriemhild knew him as her kinsman and her brother's most loyal vassal, and she trusted him. 'It is on his back, between his shoulder-blades,' she said. And she sewed in silk a little cross upon his tunic above the place. And Hagen, having learnt that which he had come to learn, left her. 'I will remember, and watch for the mark,' he said.

Then Hagen made as though he had received word that the Saxons were afraid, and had returned to

Saxony; and when Siegfried, ready for battle, came to Gunther and said, 'When do we ride forth?' Gunther said, as Hagen had bidden him, 'There will be no battle; the Saxons are fled. Let us instead ride out to hunt in the Odenwald.'

And so Siegfried rode out to the hunt with Gunther and Hagen and the lords of Burgundy; but Gernot and Giselher stayed in the king's house, for they had refused all hand in what was to come.

The hunting was successful; and in the chase, as in all else, Siegfried proved himself more skilled than any other there, so that they said to him, 'You will leave the forest empty of game, if you do not cease your slaying.' And Siegfried laughed, and he was happy.

In the afternoon they built fires in a forest clearing to roast some of the deer they had slain, and Siegfried and Gunther and Hagen and the great lords of Burgundy sat together and ate and were merry. But when Siegfried would have drunk, he found there was no wine. He turned to Gunther. 'How is this, my friend? Must we go thirsty?'

Quickly, Hagen answered for his king, 'The fault is mine, lord, for I mistook the place where we were to hunt, and have sent servants with the wine elsewhere.'

'It is a pity,' said Siegfried, 'that we are not nearer to the Rhine, for I am thirsty.'

'I know this place well,' said Hagen. 'There is a spring close by. If you would drink, come with me and I will take you there.'

'Willingly,' said Siegfried.

They rose and went, and Gunther followed them; and after him, some others of his lords. That they might be alone together, Hagen said to Siegfried, 'I have heard that few men can keep pace with you when you run. The spring is yonder. Let us run together towards it and see who runs the faster.'

Siegfried laughed. 'We are ill matched, Hagen. You have heard truly, and, besides, you are an older man than I. I will carry my sword and my shield and my hunting spear, while you run weaponless, and I will undertake to reach the spring before you.'

'Let me see you do it,' said Hagen.

So they ran together and Siegfried reached the spring first, even as he had promised. He dropped his shield and his sword, Balmung, on the grass at his feet, and leant his spear against a tree, and called out, laughing, to Hagen, coming after him.

Hagen reached him. 'You run well,' he said. He pointed to the spring. 'Drink, if you are thirsty.'

But Siegfried, looking back, said, 'Gunther is coming after us. It is fitting that the king should drink first.' And with courtesy he waited for Gunther.

Gunther knelt beside the spring and drank from his cupped hands, while Hagen and Siegfried waited. Then he rose and gave place to Siegfried, who knelt down and bent low over the gushing water.

Quickly, Hagen took up Siegfried's shield and Balmung and laid them farther away, then he took Siegfried's hunting spear from where it stood by the tree and stepped quietly behind Siegfried. On Siegfried's tunic, at the back, between his shoulder-blades, was the little cross in silk which Kriemhild had embroidered. Hagen raised the spear and, with all his strength, drove it through the cross to Siegfried's heart.

Siegfried twisted round and groped on the grass where he had laid Balmung, but did not find it there. 'False friends,' he said, 'why have you slain me? I have never done you harm. And you, Gunther, I saved your life in Eisenland and won a bride for you. Is this how you repay me?' He gasped for breath and whispered, 'Gunther, if you have any honour left, you will deal kindly with my wife. She is unfortunate, for she has murderers for her kinsmen.' His head sank forward among the flowers in the grass, and he died.

Now that it was too late, Gunther began to weep, regretting the deed; and the lords of Burgundy, coming and seeing Siegfried dead and a spear in Hagen's hand, stood silent. But Hagen said scornfully, 'Why do you weep, my king? Our trouble is ended. I am glad that it was I who slew him, since it was done for your sake.'

Gunther's lords laid Siegfried on a shield and said, 'Let us tell Kriemhild that robbers slew her husband in the forest when he was alone. So may we hide Hagen's guilt.'

But Hagen said grimly, 'I care not who knows the truth. I will take him back to Worms. Little does it matter to me if Kriemhild mourns, so long as I have done my duty to my king.'

So Hagen went back to the king's house with Siegfried's body, and there he bade his men lay it before the door of Kriemhild's chamber. And in the morning, when she came out with her maidens to go to mass, she found it, and went almost mad with grief; and none could comfort her.

Siegfried's body was set upon a bier and carried to the minster, and beside it walked Kriemhild. There, in the minster, gathered the lords and warriors of Burgundy to mourn for him; and there came Gunther, Gernot and Giselher, and Hagen. Gernot and young Giselher sought to comfort their sister, but she only wept the more. Gunther came to her. 'It is a great grief that has fallen on us,' he said. 'I shall ever mourn for Siegfried's death.'

But Kriemhild turned from him. 'You quarrelled with him,' she said. 'I have no trust in your good faith.' And she thought of how Hagen had come to her and bidden her sew a mark on Siegfried's tunic over the spot where the linden leaf had lain. She said, 'I have heard it said that when a man has been murdered, his wounds will bleed afresh when his murderer stands close by. Let me know that you are all guiltless of my husband's death. Go you all and stand beside the bier, one after the other.'

'He was slain by robbers, as I have told you, dear

sister. Nevertheless, if it will comfort you, let us do as you ask,' said Gunther. He went and stood beside the bier, and, after him, in turn, Gernot and Giselher and the lords of Burgundy; while Kriemhild watched.

And when it came to Hagen's turn the blood flowed afresh from Siegfried's wound and dripped down from the bier. Kriemhild sprang up, pointing at Hagen, and cried out, 'There stands my Siegfried's murderer!'

Gunther laid a hand upon her arm. 'Dear sister, you are mistaken, Hagen is guiltless. They were robbers who slew Siegfried in the forest.'

Kriemhild drew away from him. 'I know those robbers well. It was Hagen did it, and for your sake, Gunther. I pray that Siegfried's kinsmen may avenge the deed.' But Hagen only shrugged his shoulders and went his way, while Kriemhild wept.

And so Siegfried was buried in Worms with all honour, and many mourned for him, for he had been a good friend to the Burgundians.

2

The Vengeance of Kriemhild

After the murder of Siegfried, his father, King Siegmund, urged Kriemhild to leave Burgundy and return to Xanten, where she might dwell in honour as the widow of his son. But she would not. 'My Siegfried is buried here,' she said. 'Shall I leave him to go to the Netherlands?' And though she left her brothers' home, she remained in Worms, mourning Siegfried ceaselessly, and ever wondering how she might avenge him.

After a time she sent to Xanten for Siegfried's great store of gold which he had won in the northlands while only a youth, and she gave of it generously to all who came to serve her.

Seeing this, Hagen said to Gunther, 'Mark, my king, how your sister spends Siegfried's gold in winning warriors to her service. Soon there will be many who, for the sake of Siegfried's gold, would fight for her against whom she wills. It were best that you took the gold from her.'

'I have wronged her enough already,' said Gunther. 'She is my sister; let her keep her gold.'

'That is not wise,' said Hagen.

But Gunther would not hear him. 'Let me be, and speak no more to me of it. It would be a crime to take from Kriemhild that which is hers by right.'

'Then, once again, let mine be the crime,' said Hagen. He went with Gunther's warriors to Kriemhild's house at a time when there were few there to prevent him, and in spite of Kriemhild's anger and distress, he took away the gold in the chests and coffers where she kept it.

Gernot was indignant when he heard of Hagen's deed. 'It was ill done,' he said. And young Giselher, to whom alone of her three brothers Kriemhild still showed friendship, said, 'Had any man than our loyal Hagen acted thus, I would have slain him for it.'

But Hagen took the gold secretly and sank it in the Rhine, telling no one the place where it was hidden, save only those for whose sakes he had hidden it—Gunther and his brothers.

Now, it happened at this time that Queen Helka of Hunland died, and King Etzel's lords said to him, 'It is not fitting that you should be without a queen.' And so he sent Rüdiger, lord of Bechlaren, who was sworn to his service, into Burgundy to woo Kriemhild for him.

Rüdiger bade farewell to his wife Godelind and to his young daughter, and rode with great state to Burgundy. In Worms he was made welcome, and Gunther was glad when he learnt of his errand. 'Too long has my sister sorrowed,' he said. Gernot and

Giselher, too, rejoiced, for they loved their sister; and all the lords of Burgundy spoke of the honour done her by the great King of the Huns. And only Hagen

counselled against the marriage. 'No good can come of it,' he said.

'She has suffered enough, and thanks to you, Hagen,' said Giselher. 'Have you the heart to grudge her this comfort?'

But Kriemhild herself would have none of Etzel's wooing. 'Having been the wife of Siegfried, could I want another lord?' she said. 'And, besides, I am a Christian woman, and King Etzel is a heathen. Would you have me marry with a heathen?'

Then she thought of all she had heard of King Etzel; how rich and powerful he was, and how many brave warriors, both Christian and heathen, served him; and she thought she saw a way towards her vengeance. She sent for Rüdiger and received him kindly, asking him, 'Should I have warriors to serve me in King Etzel's land? If any did me harm, would there be men I might count upon to avenge the wrong?'

'Lady,' said Rüdiger, 'in Hunland there will be many warriors ready to serve their queen. But had you no one else in all that land save me and my kinsmen and my men, you would still be well served.'

'Swear to me, good Rüdiger,' said Kriemhild, 'that if any do me harm, you will be the first to avenge it.'

'I swear it,' replied Rüdiger.

For the first time in many months, Kriemhild smiled. 'Then I will marry King Etzel,' she said.

And so Kriemhild went into Hunland and became the wife of King Etzel; and all so that Siegfried might be avenged. And at the days-long feasting for the marriage Kriemhild looked about her at all the warriors who had come to Hunland from every country to serve the great king; and she was glad to see them so many.

When a few years were passed, Kriemhild, honoured and obeyed among the Huns, thought that the time for her vengeance was ripe. She said to Etzel, 'It is many months, my king, since I have looked upon my brothers. It would give me great joy to greet them once again. For my sake, could you not send into Burgundy and bid them to a feasting? It is a long journey from Worms, but if you ask them, I think that they will not refuse to come.'

'Willingly shall I do what will please you,' said Etzel. And he sent for his minstrels, Werbel and

Schwemmel, and ordered them to go to Burgundy and bid Gunther and his brothers and all their noble lords and warriors to a feasting at midsummer. 'Tell the queen's brothers that they will be most welcome in my land, and that their sister longs to see them once again.'

'Their lords and vassals shall also be welcome,' said Kriemhild. 'Good minstrels, see that, above all men, Hagen of Troneg fails not to come. For I have a mind to see Hagen once again.'

When Gunther heard Etzel's message, he wondered how best to act, for he had parted from Kriemhild in little friendship, yet he would willingly have been at peace with her; and, too, he did not wish to show discourtesy to so great a king as Etzel. He sent for his brothers and his lords and asked their counsel.

Gernot and Giselher were glad to think that they might see Kriemhild once again, and of all Gunther's lords only Hagen counselled against their going into Hunland as King's Etzel's guests. And when Gunther paid no heed to his words, he said to him in a low voice, 'Have you forgotten what we did? Kriemhild has a long memory. How should we be fools enough to go into Etzel's land?'

But Gunther brushed his doubts aside and spoke eagerly of the journey. 'And you shall lead us, Hagen,' he said. 'You know the way, for you were once a hostage in Hunland.'

Yet still Hagen said, 'It would not be wise to go.'

'You speak for yourself, Hagen,' said Gernot. 'For you, it may well not be wise.'

'Stay you at home, Hagen, and have a care to yourself,' jeered young Giselher. 'Let only those who are not afraid ride into Hunland.'

Hagen sprang up in anger. 'No one will be readier than I to ride with my king, wherever he leads. I will prove it, by going into Hunland, though no good will come of it.' He turned to Gunther. 'At least let us go on this death-ride well armed,' he said.

And so Werbel and Schwemmel returned to Hunland and told King Etzel that at midsummer Gunther and his lords would come; and the king was glad at the news.

'And Hagen,' asked Kriemhild, 'will he come too?'

'He will indeed.'

Kriemhild smiled. 'I am glad, for Hagen is a brave warrior. I would not have him stay at home.'

The Burgundians, all save Hagen, rode merrily enough into Hunland with Volker the minstrel to cheer their journey. On the way they passed through Bechlaren where Rüdiger, who had wooed Kriemhild for Etzel, held wide lands. He made them welcome and they remained in his house for three days. His wife, Godelind, was kindly and gracious to the strangers, and the beauty of his daughter pleased all the younger warriors.

'Were I a king,' said Volker, 'with a golden crown on my head, then should I chose your daughter, lady, to be my queen.'

Godelind was pleased, but Rüdiger said, 'My daughter could not look to wed so high as with a king.'

'But with a king's brother, maybe,' said Gunther. 'She is a fair maiden, and Giselher has as yet no wife.'

And so, with great rejoicing, Giselher and Rüdiger's daughter were betrothed. 'When we ride home again from King Etzel's feasting,' said Giselher, 'you will ride with us.' And the maiden was happy, for he was a strong and handsome young man.

Before they left his house, Rüdiger and his wife gave gifts to all their guests: to Gunther a coat of mail, to Gernot a sword, twelve golden arm rings to Volker, and an embroidered tunic to Dankwart, Hagen's brother. Hagen alone refused all gifts. 'I would have that shield hanging on your wall, or nothing,' he said. And Godelind herself fetched it down for him, and he took it and was glad. And to the Burgundians Rüdiger promised his friendship for evermore. Then he and his men rode with them for Etzel's court.

When word was brought to Etzel that they were in his land he made ready to receive them. 'Welcome them kindly, Kriemhild my wife,' he said, 'for they have come a long way to do you honour.'

'I am indeed impatient for a sight of my kinsmen,' Kriemhild said, and she ran to the window to watch for them. But in her heart she was thinking, 'Soon now, will my Siegfried be avenged. And, so long as Hagen dies, I care not who dies with him.'

Now, at that time, King Dietrich of Bern was at

Etzel's court with old Hildebrand and his three and forty faithful Amelungs, awaiting a time when it might please King Etzel to give him Hunnish warriors to fight for him against the Emperor Ermenrich, who had taken his lands from him. Dietrich was close by when word was brought to Etzel and Kriemhild of the coming of their guests, and when he saw the light in Kriemhild's eyes and watched her secret smile of triumph, he was afraid for the Burgundians. So he rode from Etzel's house that he might meet them first alone and warn Gunther. When the Burgundians knew who he was, they greeted him gladly, but he said to them gravely, 'Do you not know that Kriemhild still mourns for Siegfried?'

'Let her comfort herself with King Etzel's love. Siegfried is dead,' said Hagen.

'What if she does still mourn for him?' said Gunther. 'We are Etzel's guests and Kriemhild's kinsmen, to whom she sent loving messages to bid us come to her.'

'Siegfried is dead, but Kriemhild lives, so have a care, King Gunther,' counselled Dietrich.

Hagen smiled grimly, but said nothing, and Gunther hesitated, uncertain of himself, but Volker said, 'It is too late to turn back. Let us go on and see what awaits us among the Huns.'

So they rode on to Etzel's house with Dietrich, and Kriemhild came out to welcome them with her women and her warriors. 'I am glad of your coming,' she said. Yet she did not so much as glance at the others, but went to Giselher alone and took him by the hand and kissed him. When Hagen saw her, he narrowed his one eye, settled his helmet more firmly on his head and said to Gunther, 'I see we are not all welcome to the Queen of the Huns.'

Kriemhild overheard him and, in her hatred for him, forgot to dissemble. 'Let those who are glad to see you, welcome you, false Hagen. You are no friend of mine. What have you brought me from Worms, that I should welcome you?'

'Had I known that you wished for a gift,' replied Hagen, 'I could have brought you one. I am rich enough.' He laughed harshly.

'You could have brought me Siegfried's gold that you stole from me,' she cried furiously.

'I had enough to carry into Hunland without bearing gold for you—my helmet and my shirt of mail, a shield and this good sword.'

Then Kriemhild swallowed her anger and hid her hatred as best she might, and led Etzel's guests to him in his hall where he greeted them joyfully, for he knew nothing of Kriemhild's plotting. For Hagen he had an especial welcome, because Hagen had grown to manhood in Etzel's court. 'You have changed, Hagen, in all the years since I saw you last. Your hair is grizzled and you have lost an eye, but it does me good to see you. For I am an old man now, Hagen, but the sight of you reminds me of those days when you were only a lad and I was still a strong warrior.'

Gunther and his brothers talked with Etzel in his

hall; but Hagen went out quietly into the courtyard and beckoned Volker to go with him. 'Let us see,' he said, 'whether or not Kriemhild means harm to us.'

They sat down upon a bench outside the queen's bower, and Kriemhild looked out of a window and saw them. She called her warriors to her. 'There sits false Hagen who murdered Siegfried. I will go down and speak with him. Come with me and, when I bid you, fall upon him and slay him.' She decked herself in her royal robes and her jewels and set her crown upon her head, then she went down the outside steps from her bower to the courtyard, followed by her warriors.

Volker looked up. 'See yonder, Hagen. Here comes Kriemhild. Surely, in time of peace, for even a king's wife she is over well attended by armed men? Take care for yourself, Hagen.'

Hagen turned his head to watch them for a moment. Then he turned again to Volker. 'If they fall upon me,' he asked, 'will you stay by me, or will you go?'

'I will not go,' said Volker.

Hagen laughed. 'Then these Huns had best have a care, for there will be two of us.'

As Kriemhild came nearer, Volker made to rise. 'Sit down,' said Hagen. 'Would you have them think you are afraid?'

'She is a great queen,' said Volker. 'I would do her honour.'

'I would do her none,' said Hagen. 'But here is a sight for her to see.' And he drew the sword that was in his scabbard and laid it across his knees; and, as she came by, Kriemhild saw the green jasper flashing in its golden hilt, and she knew that the sword was Balmung. White with anger, she said, 'How dared you come here, Hagen, to the land where I am queen?'

'My three lords were bidden to Hunland. I am their vassal and have never yet refused to ride with them to any place.'

Her voice shook. 'Murderer and thief! You killed my Siegfried and you sit there holding his sword.'

'I have never denied that I killed him. I alone was guilty.'

She turned to her men. 'You have heard him admit it. Now, if you are loyal to me, avenge my husband's death.'

Hagen took a tight grasp of Balmung's hilt and Volker took hold of his weapon, and they both waited. But the Hunnish warriors looked at one another and murmured amongst themselves, 'What was this Siegfried to us, that we should risk our lives to avenge him?' And one old warrior said, 'I remember Hagen when he was a youth. Strong and grim he was, even then; and by the looks of him he is stronger and grimmer now, for all his one eye. We had best beware of him.'

'Shall we risk our lives for a woman's anger?' they said, and one by one they moved away.

When Kriemhild saw how her warriors had failed her, she returned to her bower to weep. But Hagen and Volker, having learnt her full intent, went to the hall to feast with Etzel, and to warn their king.

The Burgundians were led to a well-furnished guest chamber for the night, and there, for all the danger, they lay down to rest while Volker played sweet music to them until they slept. But Hagen armed himself and stood outside the door to watch; and when the others were asleep, Volker laid aside his harp and took up his sword instead and came out to Hagen; and so they watched all night.

The next day, before King Etzel, who was well pleased at the sight, the warriors showed their strength and skill at arms: Huns against Burgundians and Burgundians against Huns; and much honour was won by Gunther's men.

But Kriemhild, shamed and angry that her warriors had failed her, sent for Dietrich and spoke to him secretly. 'You are an outcast from your own land,' she said, 'and a guest in Etzel's house. He has shown you much kindness. If you would please his wife and have her gratitude for evermore, then kill Hagen for me. For he murdered my Siegfried.'

Dietrich was indignant. 'Hagen is no enemy of mine, that I should kill him. And he is King Etzel's guest, even as I am. The king would be ill pleased, were I to force a quarrel on him, even for your sake. No, fair queen, it is not I who can avenge your Siegfried for you.'

Then Kriemhild spoke apart with Blödel, the younger brother of King Etzel, and tempted him with promises of gifts and gold; and at last he agreed to do her bidding. 'I will slay Hagen for you,' he promised, 'and Dankwart also, lest he should avenge his brother.' And he went to choose out the bravest from amongst his men to do the deed with him.

That evening, in the king's hall, Etzel feasted with the noblest of his guests, while the Burgundian warriors, in the charge of Dankwart, feasted in the warriors' hall. And there, in the warriors' hall, unknown to their king, certain Huns, with Blödel at their head, fell upon the Burgundians. Valiantly the Burgundians defended themselves, battering at their foes with benches and tables when they had not time to reach their weapons; and many Huns were slain, Blödel amongst them. But the guests were outnumbered and their courage was of no avail, and in a little while they all lay dead amid the wreckage of the feast; all save Dankwart, who fought so strongly that none might overcome him. And his one thought was that, before he, too, was slain, he must warn his brother Hagen.

Inch by inch he fought his way out of the warriors' hall, across the courtyard to the king's hall and entered in on the mirth and revelry, his sword and his armour red with blood, and he called to Hagen, 'You have feasted too long, my brother Hagen. King Gunther's men lie dead in the warriors' hall, and all because I am your brother.'

Hagen leapt up, Balmung in his hand. 'This is Kriemhild's work,' he said. 'She wants blood: she shall have it. Guard the doorway, Dankwart. Let no Hun through alive.' And he struck down those Huns who sat nearest to him, and instantly every man was on his feet and weapons flashed in the hall; rich meats and cups of wine were flung aside and benches overturned, as guests and hosts fell upon each other. Gunther sprang down from the high table calling to his lords to lay down their weapons; but he was too late and no one paid him any heed, so he took up his sword and joined in the fighting; and Volker went to stand beside Dankwart to keep the door.

From his high seat King Etzel watched the slaughter with horror, powerless to prevent it; while Kriemhild cowered beside him, knowing it was she who had caused the strife. For though he was a great king, and had been a mighty warrior in his youth, Etzel was an old man and his fighting days were over.

While the Huns and the Burgundians fought and slew each other, to one side of the hall, apart from the fray, stood Rüdiger and the men of Bechlaren; for they were neither Huns nor Burgundians and had no quarrel with either. Dietrich of Bern, also, drew his Amelungs aside and watched, distressed.

Kriemhild saw him and screamed to him above the din and clash of swords, 'Dietrich, good Dietrich, help us from the hall or the king and I will be slain. If you were ever grateful to King Etzel, help him now.'

Dietrich sprang upon a bench and called back to her over the heads of the fighting men, 'Queen Kriemhild, how can I help you? I cannot help myself or my own men. And no one can leave the hall while Dankwart and Volker guard the door.'

Kriemhild wrung her hands. 'Dietrich, have pity on us, for you are our only hope.'

'I will do the best I can, Queen Kriemhild.' Dietrich stood upon a table and cried out with a mighty voice, bidding the strife to cease. Gunther heard him and called out, 'That is King Dietrich's voice. Pray God that in our anger we have not slain any of his men, for he is our good friend.' And he stood upon the table beside Dietrich and shouted for the fighting to cease until he had heard what Dietrich had to say. And because both Huns and Burgundians alike respected and honoured Dietrich, they lowered their swords and waited.

'Good King Gunther,' said Dietrich, 'let my men go safely from the hall, for we are Amelungs and have no part in your quarrel with the Huns.' And the Amelungs echoed his words; all save the youth Wolfhart, Hildebrand's nephew, who was rash and hot tempered and would willingly have joined in the fighting. 'Why ask favours, my king?' he cried. 'With our swords we can win past Dankwart and the minstrel.'

But Dietrich silenced him. 'Hold your peace Would you fight with our friends?'

'Willingly,' said Gunther, 'do I give you safe conduct from the hall, good Dietrich. Take with you your Amelungs and any friends you please.'

Dietrich went to the king's table and took King Etzel by the hand, and with the other arm he raised the trembling Kriemhild and led them from the hall; and after him went Hildebrand and the Amelungs, all thankful to leave safely save Wolfhart, whose looks were sullen as he fingered his sword hilt.

Then Rüdiger spoke up, 'Let me and the men of Bechlaren go in peace also, King Gunther.'

But before Gunther could reply, young Giselher called out, 'Go, good father of my bride, for there is only peace and love between us. You have been a true friend to us.'

So Rüdiger and the men of Bechlaren went from the hall, and they and the Amelungs went to their own separate lodgings in the king's house and thought to have no part in the fray. But Etzel bewailed the loss of his men within the hall to those Huns who crowded about him in the courtyard. 'Alas,' he said, 'this is surely the worst feasting that ever any king gave.'

In the king's hall the fight started anew, and did not cease until all the Huns lay slain; and then those Burgundians who remained alive laid down their swords and looked about them to see which of their friends still lived. And Gunther and his two brothers, Hagen and Dankwart, and Volker the minstrel were among those who had survived.

In the morning Etzel came out to the courtyard

with his men and stood before the king's hall to see what had chanced in the night, and Hagen and Volker came out through the doorway to the top of the steps which led down from the hall to the courtyard.

'Too many of my good warriors have you slain,' said Etzel. 'Ill guests have you proved yourselves.'

Hagen laughed mockingly. 'To me, at least, King Etzel, you should show only friendship; for had I not killed Siegfried, fair Kriemhild would not now be your wife.'

Kriemhild heard his words and was enraged that he dared flout her before the Hunnish warriors, and she said, 'I will give the king's great shield, filled with gold, to the man who brings me Hagen's head.' And many, hearing her, tried to fight their way up the steps to the king's hall to do her will. But all who tried were slain.

Yet, as the long midsummer day wore on to its ending, the Burgundians could no longer rejoice at their victories. 'Here in the king's hall we are safe,' they said. 'The Huns cannot come in to us; but nor can we win out to freedom, for they stand close around and they outnumber us by hundreds. We shall never see Worms or the Rhine again. Yet it would be better to die out there, in the open, by the swords of our enemies, than to perish here slowly, like rats in a trap. Let us send to King Etzel and ask him to let us come forth and die quickly.'

So they called to Etzel to come and speak with them. Weary and angry, the old king stood in the courtyard below. 'What do you want of me, men of

Burgundy? You were my guests, and you have slain my noblest lords and many of my finest warriors. I cannot believe that you would dare to ask for peace.'

'The wrong you did us is as great as the wrong we have done you,' said Gunther. 'For it was the Huns who struck the first blow. We do not ask for peace, King Etzel. We know our lives are forfeit. The only favour we would beg of you is that you bid your men stand back from the hall and let us come out to you. Then tell your warriors to fall upon us quickly, for we have fought long and we are tired and we would rest. Grant us this one mercy.'

Etzel admired their courage, and, for all his anger and grief, he would have granted their request; but before he could speak Kriemhild cried out, 'Beware of letting them come down to you, men of Hunland. For my brothers are strong and mighty, and they will slay many of you before they themselves lie dead.'

'Sweet sister,' Giselher cried to her, 'we have ever loved each other and been good friends. Why did you bid us come to Hunland to all this grief? We came because we trusted you. You must have pity on us now.'

'What pity did you have, Giselher, you and Gernot, when you did not prevent Hagen from slaying my Siegfried? Yet, in the name of the king, who will grant what I ask, I promise you, my brothers, you and all your men, your lives and freedom to return in peace to Worms, if you will give me Hagen. Give me Hagen, bound, and the rest of you may go.'

But her brothers all cried out against her and said, 'Hagen is our man and has served us loyally. Though we were a thousand others instead of so few as we are, we would not buy our lives at such a price.'

And in her rage that she was once more cheated of her vengeance, Kriemhild called to the Hunnish warriors to set fire to the king's hall and stand close about it, that none might come out to escape the flames.

It was done as she had bidden, and all night the great hall of King Etzel burnt, so that no one watching would have believed any was left alive in it. Yet the arches of the vaulted roof were strong and held, and though the thatch burnt, the roof did not fall in; and by dawn the fire had burnt itself out and there were yet some among the Burgundians who still lived.

Eagerly Kriemhild's warriors hastened to the hall, each wishing to be the first to find Hagen's body and tell her of it and so earn a reward. But as they rashly crowded into the hall where they thought to find only dead men lying, they found Gunther standing, and Gernot and Giselher, Dankwart, and Volker the minstrel with Hagen by his side.

'We are still here,' said Volker. 'You are welcome.' And he flung a spear and slew the foremost of them; and then he and the others fell upon the Huns so that those who were not killed were forced to flee.

When the sun was higher, Rüdiger came from his lodging to see how the Burgundians had fared in the fire, for he longed that they might yet make peace with the king. While he stood in the courtyard with a heavy heart, a Hunnish warrior said, 'See how he weeps for the king's enemies, the man whom our good King Etzel has raised above all others. Is that how he shows his loyalty?'

Kriemhild came to him and said, 'Do you remember, Rüdiger, how you were my first friend in Hunland? When you came wooing for King Etzel you

swore that if ever I had need of you, you would serve me. Today I hold you to that promise. Go and kill Hagen for me.'

'I promised to give my life for you, but not my honour. I brought those men into Hunland to a feasting. I could not kill one of them.'

Kriemhild knelt before him and wept. 'You swore an oath to me, Rüdiger.'

'And I swore friendship to your brothers.' He turned away from her, but Etzel said to him, 'I have given you land and gold and honours and loved and trusted you above all my men, though you are no Hun. Will you fail me when I need you?'

'Take back all the gifts you gave me, my king. Let me go forth into the world with my wife and daughter, penniless and in rags. Acquit me of my vows to you. For I took those men into my house and gave them gifts and promised friendship to them. And I betrothed my daughter to Giselher.'

'Your oaths to me came earlier,' said Etzel. 'And I have no one else left of any worth.'

Rüdiger turned slowly away and went to his men. 'Arm yourselves,' he said, 'for we go to fight against King Gunther.'

The men of Bechlaren, with Rüdiger at their head, went towards the king's hall, and Volker saw them come. 'Here come more warriors against us,' he said wearily. 'Would God it were all over.'

But Giselher cried out, 'It is Rüdiger. He has come to fight for us, for the sake of his daughter, my bride.'

Rüdiger stood before the hall and called out, 'Defend yourselves, brave men of Burgundy. I would have helped you, but I must fight against you.'

'I cannot believe,' said Gunther, 'that you would be false to us.'

'I may not do otherwise,' replied Rüdiger.

'You gave me a fine sword as a gift, Rüdiger,' said Gernot. 'I would not slay you with it.'

'I would to God, friend Gernot, that I were dead and you were safe home in Worms.'

Giselher pleaded with him, 'Would you make your daughter a widow before she is a wife?'

'If by God's grace you escape with your life, good Giselher, do not let my daughter suffer because her father failed you.'

Hagen held up his battered shield. 'It was a good gift you gave me in your house at Bechlaren. But it is useless now. I would be glad of another such gift today.'

Rüdiger stepped forward and held out his own shield. 'Take it, friend Hagen. It is the last gift I shall ever give. I wish it might see you safely home to the Rhine.'

Hagen came down the steps from the hall to take the shield. 'I will give you a gift in return,' he said, 'for you are a good man. Against you alone I will not fight today, though you slay all the men of Burgundy before my eyes.'

At that Volker called out, 'And I also will not fight with you, good Rüdiger, for the sake of your kindness.

See, I am wearing the arm rings which your wife gave to me. If you come safely home, tell her that I wore them at King Etzel's feasting, even as I promised her.' And he held up his arm that Rüdiger might see the sun glitter on the gold.

'I will tell her, should I see her again,' said Rüdiger.

After that they spoke no more, but fought. The Burgundians stood aside and let the men of Bechlaren into the hall and fell upon them there; and in the fighting Gernot and Rüdiger died by each other's hand, so that Rüdiger was slain by his own gift. When the other Burgundians saw that Gernot was dead, they fell fiercely upon Rüdiger's men, and Hagen and Volker, who had sworn to spare only the one man among their adversaries, did great deeds; and in a little while all the men of Bechlaren were dead. But the Burgundians wept for Gernot and for Rüdiger their friend.

When Rüdiger and his men did not return to Etzel, Kriemhild said, 'He is false and has cheated us,' and she went across the courtyard to the king's hall and stood at the foot of the steps. 'Why does Rüdiger tarry?' she called. 'He is a traitor if he has made peace with our enemies.'

'You do him wrong,' said Volker. 'He served you loyally.' And they brought his body out to show her; and Kriemhild wept that Siegfried was still unavenged.

But Etzel and the Huns mourned for Rüdiger, who had been the noblest of all King Etzel's vassals. The sound of their lamentation reached to Dietrich where he sat in his lodging with his men, and he wondered at it. Wolfhart sprang up eagerly. 'I will ask what has happened, my king.'

'Not you,' said Dietrich, 'for you are headstrong and too fond of quarrels. Let another go.'

So another of his men was sent to find what had befallen and he returned weeping. 'Good Rüdiger is dead. The Burgundians have slain him.'

'That was ill done,' said Dietrich. 'He was their friend. Yet I cannot believe that they would be so base. Hildebrand, go you to King Gunther and ask further of the matter.'

Hildebrand would have gone unarmed, but Wolfhart said, 'If they were indeed treacherous towards Rüdiger, my uncle, how will you fare at their hands? It were best to go armed and ready for the worst.'

So Hildebrand armed himself and, while he made ready, all the three and forty Amelungs made ready also. 'I would go alone,' he said to them. But they replied, 'We would stand near by to guard you in case of treachery.' And Hildebrand let them have their way.

He went before the king's hall and spoke with Hagen and asked if the tidings were true, and Hagen told him how Rüdiger had died.

'It is a sad day,' said Hildebrand.

'Willingly would I fight with you and avenge good Rüdiger,' cried Wolfhart, 'but King Dietrich has forbidden it.'

'The king has forbidden it. A fine excuse for cowardice,' laughed Volker.

Wolfhart was so enraged that he rushed forward against Volker and all the men of Bern followed him, and even Hildebrand went with them when he saw that he could not hold them back. On the steps they fought, and in the hall, and there Volker was slain by Hildebrand, and Dankwart was slain; and the two young men, Giselher and Wolfhart, fell by each other's hand.

When Hildebrand saw that Wolfhart was wounded, he went to him that he might carry him from the hall, but Wolfhart said, 'Good uncle, save your strength to fight against Hagen, for he is to be feared. There is no more that you can do for me. But when anyone would weep for me after today, say that I need no tears, for I died bravely. It was a good fight, my uncle.' Wolfhart smiled a little, then he died.

When Hagen saw that Volker was slain, he sought out Hildebrand that he might avenge his comrade; but Hildebrand thought, 'Where is the use in being slain in the quarrels of others? I am Dietrich's man. I have loved and served Dietrich since the day he was first sent to me as a child. For him only am I ready to die. Now that his few faithful Amelungs are all dead, shall I be slain also and leave him in exile alone?' He went swiftly from the hall and ran to Dietrich.

When Dietrich saw him come, in his armour and wounded, he said, 'There has been fighting. And I forbade it. Is Rüdiger indeed dead?'

'He is dead, and I have hardly escaped with my life,' replied Hildebrand.

Dietrich rose. 'I will go myself to talk with the men of Burgundy. Tell the Amelungs to arm themselves and come with me.'

'There is no one to tell,' said Hildebrand. 'Your only man stands here. All the others are slain by the Burgundians.'

'I thought,' said Dietrich slowly, 'that the worst that could befall me had already happened, but now I see that I was wrong. Once I was a king, with warriors to serve me and lands to rule, and now I have but one friend left.' He fell silent for a time, then he looked at Hildebrand. 'Yet am I not utterly comfortless, for the friend that is left to me is the best one of all. Come, Hildebrand, let us go together and see if we can end this tragic strife.'

Dietrich armed himself and went with Hildebrand to the king's hall, and found that, of all the Burgundians, only Gunther and Hagen were left alive.

'Here comes Dietrich to avenge his men,' said Hagen.

'You are brave warriors,' said Dietrich. 'For all you have slain my Amelungs, if you will yield to me, I will see that you go safely home to Burgundy. Etzel, I know, will grant me the boon.'

'We have fought for many long hours,' said Gunther, 'we shall fight to the end.' And he and Hagen came down the steps and fell upon Dietrich and Hildebrand. But though they fought valiantly, even as they had fought all day, they were weary and wounded and had little strength left; and because Hildebrand

and Dietrich thought to be merciful, they were taken alive.

Dietrich bound them and led them to Kriemhild, thinking that as one was her brother and the other a kinsman, she would forget her anger and forgive them the wrong they had done her.

'I will ever be grateful to you, Dietrich, for your service to me today,' she said.

'Then, good queen, let these brave men live for my sake.'

'I will do it gladly,' she said. And he believed her, because he was without guile himself, even as Siegfried had been.

She had Gunther and Hagen laid in chains, apart from each other; and Balmung she took from Hagen and kept for herself. Then she went alone to Hagen. 'Give me the gold you stole from me, and you and Gunther may go home to Worms.'

'You waste your words,' said Hagen. 'The gold I took for my lords. While one of them yet lives, no one else shall have it.'

Angrily she left him and bade her warriors slay her brother. They went to where Gunther lay and cut off his head and brought it to her. She took the head by the hair and carried it to Hagen and dropped it down beside him. 'There is your king,' she said.

Hagen turned his head away so that she should not see his tears. Then after a little while he looked again at her and said, 'It has ended even as I thought it would, our death-ride into Hunland. Gernot is dead, and young Giselher, and Gunther my king. And as for the gold: now only God and I know where it is hidden. You shall never see it.'

In her fury and her hatred she sent for Balmung. 'This, at least, of all Siegfried's treasure, I have,' she said. She drew it from its scabbard and the green jasper glinted. With all her strength, she raised it high above her head in both her hands and struck, again and again, until Hagen was dead. And so, at last, Siegfried was avenged.

When King Etzel heard of it, he said, 'He was the greatest warrior of them all, and now he is slain by a woman. Let me not look on her again.' And he turned from her with loathing.

Old Hildebrand rose quietly and took up his sword and, before ever she knew what he was about, he cut off Kriemhild's head. 'It is better so,' he said.

FOLK-TALES

I

Karl the Great and the Robber

IN the time of the good Emperor Karl the Great, whom the Franks called Charlemagne, there lived near Ingelheim a knight named Elbegast. He was skilled in all a warrior's feats and there was no knight braver; and above all men he loved and honoured the emperor. But as much as he loved and honoured the Emperor Karl for his courage and his justice, he hated all men, whether churchmen or lords, who lived only to serve their own ends and who oppressed those weaker than themselves. With all such men he considered himself at war, and he preyed on them continually, assaulting their castles whenever he might, or attacking them as they rode abroad, and taking from them great booty. Part of these spoils he kept for himself and his men, but part he gave to the poor or to those whom others had wronged, so that he was much loved by the poor and the wretched.

In time, word of his robberies was brought to the Emperor Karl and, angry to think that any knight should set himself above the emperor's justice, he declared him an outlaw; so that it became each man's right and his duty to kill Elbegast if ever he chanced to meet with him. From that moment every man was against the robber, save a few of his more devoted followers, so that every day he and these few went in peril of their lives.

Now, at a time when the Emperor Karl was holding his court at Ingelheim, one night he dreamt a strange dream. In his dream it seemed to him that as he lay on his bed, an angel came down and stood at his bedside and spoke to him. 'Go,' said the angel, 'in the name of the Lord, and steal your neighbour's goods.' With this the angel vanished and the emperor awoke, puzzling at his strange dream.

'That an angel should bid a man steal from his neighbour is a curious thought. How odd our dreams are sometimes.' And he smiled to himself in the darkness and fell asleep again.

But once more he dreamt that he saw the angel standing beside his bed, and once more the angel said to him, 'Go, in the name of the Lord, and steal your neighbour's goods. Fail to obey this heavenly command, and you will lose both your throne and your life.'

The Emperor Karl awoke, marvelling that he should twice have dreamt the same dream. 'Truly,' he thought, 'were it not no more than a dream, I would say it to be a temptation from the Evil One, for surely no angel would counsel such wickedness.'

He thought long on his dream before he fell asleep again, but no sooner was he asleep than he dreamt the same dream yet once more. But this time the angel spoke in stern and angry tones. 'Fail to obey this heavenly command,' the angel said, 'and you will lose not only your throne and your life, but your immortal soul as well.'

A third time the Emperor Karl awoke, and this time he was perturbed. 'The same dream,' he thought, 'three times. Surely it must be more than a dream. And strange as the command is, if it comes from heaven, I would do well to obey it.'

So he rose quietly and dressed and armed himself, and taking sword and shield and lance, he went from his bedchamber and through the silent palace, taking care to wake no one as he went. 'As though I were an intruder in my own palace,' he thought; and wondered what explanation of his conduct, so unfitting to the dignity of an emperor, he could give if anyone awoke and saw him. But by this time he never doubted that he must, at all costs, obey the angel's command.

By luck, he managed to reach the royal stables unobserved, and with a sigh of relief he went to the stall of his favourite horse. He saddled and bridled the horse and set off, the first peril—that to his royal dignity—safely passed.

But now he faced the far greater peril that every robber must face—the peril of death or capture. And warily he rode along, glancing all the time from side to side, and he found himself—he who was ordinarily without fear—peering at every shadow and questioning whether every bush might not conceal a watcher. 'Truly,' he thought, 'the life of a robber cannot be an easy one. Though he may not deserve his plunder, there is no doubt that he earns it.' And he fell to thinking of the robber knights whom he had captured and hanged for their crimes, and those, like Elbegast, whom he had not caught, but who were outlaws living in daily danger and dread; and where before he had only thought of such men with cold, impartial justice, he now felt understanding and even pity for them.

For a moment he smiled to himself a little wryly, 'I would I had one such man with me now, to teach me how to rob. For otherwise I fear that long before I have stolen as much as one gold coin, I shall have blundered and been caught.' And at the thought of capture, he smiled no more. For capture, at the best, meant recognition and ridicule; while, at the worst, he might be slain before he had time to prove himself the emperor.

As he had no idea in which direction he should ride, he had given his horse its head, hoping that heaven, whose commands he was obeying, might in that way give him a sign or help him in his quest; and the horse had made for the forest.

After a mile or two in which he had seen no other living creature, the Emperor Karl suddenly saw a horseman, who rode out from among a group of trees a little way off and made towards him. The stranger's horse was black, and in the darkness, his features, like the emperor's, were hidden. The emperor called to him, 'You who ride so late, who are you and what is your errand?'

But the stranger would give him no answer. 'Then prepare to defend yourself,' said the Emperor Karl, and he levelled his lance and rode at him.

The two of them fought mightily, on horseback and on foot; and the stranger showed himself to be a worthy opponent for even one as skilled as the Emperor Karl. But at last he was overcome and he yielded. 'I have never fought with a better man,' he said. 'My life is yours. Take it if you will.'

The Emperor Karl sheathed his sword. 'What should I want with your life?' he asked. 'I would rather have your friendship, for you are a brave knight. But tell me your name.'

'I am Elbegast,' replied the stranger.

The Emperor Karl was well pleased at this. 'So heaven has sent me help,' he thought. Aloud he said to Elbegast, 'Since you, too, are a robber, let us ride together. Between us, we may find some plunder this night.'

'Willingly,' said Elbegast. 'The two of us should be a match for any who would seek to prevent us.'

They mounted their horses, and, to try his companion, the Emperor Karl said, 'What say you, my friend, shall we ride to Ingelheim and rob the emperor's treasury? Two bold men such as we are might well be successful even there.'

But Elbegast shook his head and frowned in the darkness. 'No. I may be an outlaw, but I am true to the emperor, for he is good and brave and just. Never yet have I wronged him; I shall not do so today.' He paused for a moment and then went on, 'But I know a man well worth robbing, one who deserves to lose a few of his possessions. Let us go to the castle of Count Eggerich and take what we can from him.'

The Emperor Karl was surprised at his words, for Count Eggerich was one of his noblest lords and a trusted counsellor, and married, besides, to his own sister. 'You declare yourself loyal to the emperor,' he said, 'and yet you would steal from the emperor's friend, who is married to the emperor's sister. A strange way of showing your loyalty.'

'My loyalty, such as it is, is greater than Count Eggerich's. I would stake my life on that,' said Elbegast. 'And as for the emperor's sister, I have heard it said that the count is none too kind to her, poor lady, which I can well believe.'

The Emperor Karl was astonished at this, since he had never had anything but smooth words from Count Eggerich, who, too, always spoke with respect and affection of his wife; and he was very curious to learn more of the matter. 'In that case, by all means let us go to Eggerich's castle and rob him,' he said.

They rode for the castle; and within sight of it, they tethered their horses and went the rest of the way on foot. By the skill and daring of Elbegast, they entered the castle unseen by the guards; though, had the Emperor Karl been alone, he would, he realized, have been captured very soon.

Though Elbegast wondered at the clumsiness and inexperience of his companion, who had called himself a robber, he said nothing of it. But, once inside the castle, he thought it best to face the greatest danger of their enterprise alone. 'Wait here for me, below the wall, guarding our weapons,' he whispered, 'and I will return to you with the booty and then you must help me carry it away.'

The Emperor Karl waited in the shadows, his sword held ready and his eyes peering into the darkness, starting at every sound, while the minutes seemed like hours. But the waiting was over at last, and Elbegast was beside him, staggering beneath the weight of bags of gold, silver cups and other treasures.

Elbegast chuckled softly and whispered, 'We have done well tonight.'

'Let us make speed to get away,' said the Emperor Karl. 'You must be skilled above all robbers, but even your skill will be useless if our luck does not hold. With all this booty to carry, we shall be very easily taken if we are seen.'

But Elbegast hesitated. 'I have heard,' he said, 'that Count Eggerich owns a wondrous golden horse trapping, all hung with little golden bells. He keeps it in his bedchamber. If I could steal it, that would indeed be a test of my skill, and a sad loss to Eggerich besides.

If you will wait here for me a little longer, my friend, I will go back and fetch it.'

Caution would have seemed cowardly, so the Emperor Karl said, 'Go, but for the love of heaven make all the speed you can. I will be waiting here for you, whatever befalls.'

Elbegast slipped away, and hurrying silently through the castle, he made his way unseen and unheard to the bedchamber of the count. One dim light burnt in the room, and Elbegast saw that the count and his lady lay asleep in their huge bed. He made his way to the golden caparison and took hold of it. But as he would have moved away with it towards the door, his foot stumbled and all the little golden bells tinkled clearly.

Count Eggerich started up from his sleep. 'Give me my sword!' he cried out. 'We are betrayed!'

Elbegast set down the caparison and slipped quickly into the shadows, hardly daring to breathe.

The count's cry awoke his lady. 'What ails you, lord?' she asked.

'There is someone in the chamber. I heard a sound.'

'It will be no more than the wind outside, lord. Or you have had an evil dream. You have been much troubled by evil dreams lately. There is some matter on your mind which will not let you rest, I think. Will you not tell me what it is, that I may perhaps lighten it?'

Count Eggerich laughed harshly. 'There is no need for you to fret yourself. After tonight there will be no more evil dreams. Tomorrow the matter will be ended, and I shall be master of half the emperor's lands.'

'Master of half the emperor's lands! What can you mean, lord? I know that tomorrow you ride to Ingelheim, to the emperor's council. He loves you well, but why should he give you half his lands?'

Eggerich laughed again. 'He will give me nothing. I shall take what I want. I and others of my good friends'—and boastfully he named them—'have sworn that tomorrow we go armed to the council, and there we will kill the emperor. Once he is dead no one will dare to oppose us, and his lands will be ours to share amongst ourselves.'

The lady gave a cry. 'Never will you murder my brother while I live to warn him.' She made as though to rise from the bed, but the count took hold of her and struck her so cruelly that she fell senseless to the floor. He gave one more glance around the dimly lighted room, then lay back upon the pillows and settled himself to sleep once again.

Elbegast, appalled at what he had seen and heard, would have avenged the cowardly blow, but he remembered his companion, waiting below. 'If I am slain and do not return to him,' he thought, 'he will still be waiting for me, as he promised, when daylight comes, and he will be taken and hanged.'

So, after a while, keeping still to the shadows, he made his way from the room and hurried back to where the Emperor Karl waited. He told him all he had learnt, and to the emperor it seemed almost unbelievable that Eggerich could be a traitor, yet he saw no reason to doubt Elbegast's word, and the robber's indignation was real enough. 'Give me my sword,' said Elbegast, 'and I will go back and kill this villain. If I do not return by dawn, you will know that I have failed and am dead.'

'Are you mad,' exclaimed the Emperor Karl, 'that you will risk your life for the emperor? What has he ever done for you, that you should care for his fate?'

'Were you not my friend and my companion in this night's work, and had you not spared my life in the forest, I would kill you for those words,' said Elbegast. 'May God keep the emperor from all hurt, for he is a good man. Now give me my sword and let me go back.'

The Emperor Karl laid a hand upon his arm. 'No. Let us rather go and warn the emperor. Let him deal as is fitting with Eggerich and his fellow traitors.'

'Why should the emperor take my word against that of his friend and the husband of his sister? I am a robber and an outlaw. The emperor will hang me as soon as he sees me, not listen to my talk of plots.'

'I am no outlaw,' said the Emperor Karl. 'I will be the one to warn him.'

And so it was arranged between them that he should go to Ingelheim and warn the emperor early in the morning, and that, later that day, he would meet Elbegast at a certain spot in the forest and tell him what had befallen.

The Emperor Karl returned to his palace before daybreak, and entered as he had left, unseen. That morning he called to him early, before the time appointed for the council, those of his lords and counsellors who had not been named by Eggerich. To them he told of his strange dream and how he had obeyed the angel's command, of his meeting with Elbegast the robber, and of the plot of which Elbegast had learnt. 'Surely, my friends,' he said, 'this warning came to me in my dream from heaven itself; for truly, had I not obeyed the angel's command, the wicked scheming of the traitor Eggerich would not have come to light until it was too late, and I would have lost both my throne and my life, even as the angel warned me.'

The loyal lords and counsellors were loud in condemnation of Eggerich and his accomplices, and all of them swore to stand beside the emperor and fight or die for him that day.

'There will, I trust, be no need for that, my friends. For, as each of the traitors comes to the council, he and his men will be disarmed and led to prison. They who thought to take us in their trap, they, suspecting nothing, will themselves walk into our trap.'

And so it was. One after another, the lords who supported Eggerich in his attempt, together with their men, were surrounded by the emperor's knights as they entered the palace, and before they realized their peril, it was too late and they were disarmed.

The last to come was Eggerich himself. The captive of those very knights whom he had hoped to kill or terrify into accepting his leadership, he called to his men, but found that they, too, were prisoners. He was taken before the Emperor Karl. 'What means this outrage, lord emperor?' he asked. 'Is this the way you greet a loyal vassal and the husband of your sister when he comes to your council?'

'No,' replied the Emperor Karl. 'Rather is it the way I greet a traitor.'

'I am no traitor,' said Eggerich, 'and I will prove it with my sword against any who dares to accuse me.' He faced the Emperor Karl defiantly. 'Call out your champion, lord emperor, and let him make good your charge, if he can.'

There were many there who would have offered themselves to do battle with Eggerich for the sake of their emperor, but he said, 'I have chosen my champion already.' And he sent a messenger to the forest, to the place where he had arranged to meet Elbegast, to bid Elbegast come to court with no delay.

When Elbegast saw, instead of his comrade of the night's adventure, one of the emperor's knights riding towards him, and when he heard the emperor's command, for a brief moment he hesitated, wondering if perhaps his companion were held prisoner and this were no more than a trap to take him also. But he instantly brushed his doubts aside. 'Such is not the emperor's way,' he thought, and he rode with the messenger to Ingelheim.

In the palace, before the Emperor Karl, Elbegast knelt with bowed head. 'I am a robber and an outlaw, lord emperor,' he said. 'Yet because you bade me I have come to you freely, trusting in your good faith.'

'If I did not take your life last night, when you yielded to me, why should I take it today?' asked the Emperor Karl.

At the sound of his voice, Elbegast looked up, amazed. 'It was you,' he said. 'It was you, last night in the forest.'

The Emperor Karl smiled. 'It was I. And last night you did me a service, teaching me how to rob. Today I would ask another service of you.'

'Whatever it is,' said Elbegast, 'I will do it for you gladly.'

'Will you do battle to the death for me against the traitor Eggerich?'

Elbegast sprang up eagerly. 'Most willingly, lord emperor.'

So Elbegast and Count Eggerich met in combat, and they were well matched; and a hard battle they had, and long. But in the end the right triumphed, and Elbegast was victorious and Eggerich was slain, to the great rejoicing of all those who were loyal to the Emperor Karl.

The one-time outlaw and the emperor who had been his companion in a robbery became firm friends; and Karl the Great gave to Elbegast for himself not only the wide lands of the dead traitor Eggerich, but also the hand of his widow in marriage.

II

The Mousetower

There once lived a rich, proud churchman named Hatto, a man unfitted to preach God's love in the smallest church, let alone be the great bishop of Mainz. Many tales have been told of Bishop Hatto's treachery and cruelty, but the best known is the tale of his terrible end.

In Bishop Hatto's time there came a great famine to Germany, so that the poor people starved in their hundreds; while even the rich went hungry. Some among the rich folk shared with the poor the corn stored in their granaries—but not Bishop Hatto. When men fell dead in the streets of Mainz and the little children wept for food, he only shrugged his shoulders and cared nothing for it. 'There are too many poor folk in our city, we can well spare a few of them,' he said.

As the days passed and the famine grew worse, the

townsfolk came to the bishop's house to implore his help. He considered for a while and then he said, with a smile, 'Let all who are starving gather in my large barn which stands beyond the city gates, and I can promise them that they will not go hungry any longer.'

Joyfully the people hastened to the barn and crowded inside, each struggling with his neighbour to be first; and there they eagerly awaited the coming of the bishop, believing that he would give to each one of them a share of his own hoarded grain.

But when Bishop Hatto came with his attendants, he ordered the attendants to close and lock the doors of the barn. Then he had them set fire to the building with all the folk inside. And there in the great barn they were all burnt to death, screaming to the bishop for mercy and crying for the doors to be opened.

All that Bishop Hatto said as he watched the flames was, 'The poor, like mice, eat up our food. We are well rid of these vermin.' And all that he said when he heard their cries was, 'Hark at the mice squeaking!' and he laughed.

When the burning was over and all the folk were dead, Bishop Hatto returned to his home. But that night there came, from nowhere, it seemed, a great crowd of mice to the bishop's house. They overran every room, gnawing and nibbling everywhere; food, furniture, hangings, nothing they left; and they gathered thickest in the bishop's bedchamber. Nothing would drive them away, and day and night they crowded about Bishop Hatto, squeaking and pushing and jostling. Waking or sleeping they never left him; whether he sat at his table or knelt to pray, the mice were always there. They followed him when he rode abroad; even into the church they followed him; until at last he could bear it no longer.

Now, in the Rhine, opposite the town of Bingen, stands a tall rock, and the bishop said, 'Build me a high tower of stone upon that rock, and I will go and live there, away from these mice.'

It was done as he ordered, and the bishop was rowed in a little boat to the tower on the rock, and there, locking the doors of the tower, he felt himself secure at last.

But that night, as Bishop Hatto lay abed in his tower, the mice in their hundreds swam across the Rhine. They climbed up the tall rock and over the high walls of the tower, they dropped through the window slits, they gnawed their way through the doors. And there, in his Mousetower, Bishop Hatto was eaten alive by the mice.

III

The Water-sprite and the Bear

In a mill beside a stream, a short way from a village, lived a miller. He was a cheerful, good-natured man, and would have been contented enough with his lot had it not been for one misfortune. In the stream close by the mill lived a water-sprite, a sly, ugly creature with dank hair like water-weeds, sharp, pointed teeth and flat, webbed feet.

At first it was bad enough when the water-sprite's head would suddenly appear above the water and he would look inquisitively as anyone passed by; but it was worse when he began to climb on to the bank of the stream and sit there, showing his long teeth in a thoughtful grin and watching with his unblinking, pebble-like eyes the miller or his wife, the serving-wench or the boy who helped in the mill.

'He gives me the creeps,' said the miller's wife, 'sitting there like that.' And the boy said, 'He is ugly, and no mistake.' But the serving-wench just gave a scream whenever she caught sight of him, picked up her skirts and ran.

Yet all that was nothing to what came later. The miller had had little to grumble about so far. But one day the water-sprite padded up the steps to the mill, put his head round the door, said, 'Good morning, miller,' in his wet voice and came in and settled himself comfortably on the floor in a corner by the hearth.

The miller was not one to grudge a warm, dry corner to anyone, not even to a water-sprite—the last kind of creature, surely, whom one would expect to want such a thing—but the water-sprite took to coming into the mill whenever it pleased him, at any hour of the day or night, and prowling around for all the world as though it were his own place, so that, just when one least expected him, there he would be: under the table at supper time, waiting in the kitchen the first thing in the morning, padding behind one on his silent feet when one had supposed oneself to be alone, or appearing suddenly at a dark turn of the stairs just as one was going to bed. Or perhaps the very worst of all was when, after an evening in which he had—most happily—failed to appear at all, he would be found curled up asleep in the middle of the bed, lying on a patch of damp bedclothes, for, of course, he always dripped stream water wherever he went.

In a few weeks he was quite at home in the mill, and one could be sure of meeting him there at least once every day. By this time the serving-wench had left—running all the way home to the village one night after finding the water-sprite on the stairs in the dark—and her place had been taken by no fewer than three others, each of whom had, in turn, quickly followed her example, and the miller's wife was having to do all her own work.

One evening, the water-sprite showed great interest while the miller's wife was cooking the supper, coming quite close and sniffing at the roasting meat. The next morning he arrived at breakfast time, carrying a fish on the end of a stick. He sat himself down by the hearth and broiled his fish over the fire. When it was cooked, he tasted it cautiously, liked what he tasted and ate it up in two bites. After that he always cooked his meals at the hearth, four or five fish at breakfast and supper, and scrunched them up, heads and tails and bones and all, with his long teeth; watching the miller and his wife thoughtfully as he did so, in a manner which they found most disconcerting. It quite put them off their own meals.

Another pretty trick of his was to set the mill wheel racing in the middle of the night, so that the miller and his wife would wake up in a fright, and the miller would have to get up and go to see what was the matter.

It was, of course, inevitable, with such a state of affairs, that there should come a day when the miller found himself alone in the mill. Not a single girl from the village would come to work for him, his boy's father had found the lad another master, and his wife had gone home to her mother. So the miller was all alone—all alone, that is, save for the water-sprite. But because the mill was his livelihood and had been his home as well for all his life, he had to stay on, in company with the water-sprite; and little comfort he found in such a companion.

One evening, just after dark, a bearward knocked on the door of the mill and asked lodging for the night. He was on his way from one village to another with his dancing bear. The miller sighed, for he dearly liked a good evening's talk and he saw little enough of other folk these days, and the bearward looked a cheerful fellow with a merry grin and a bright eye, a man, indeed, after the miller's own heart.

'It is unlike me to be inhospitable,' said the miller. 'Not so long ago I never dreamt there would come a day when I would turn a stranger from my door at night. The village is only a few miles on, my friend, you had best go there, to the inn.' And he told the bearward about the water-sprite. 'It is all I can do, to stay here myself,' he said. 'A hundred times a week I tell myself, "Tomorrow I shall lock the door and throw the key in the stream and go." But I always manage just one day more. Yet one cannot expect a stranger to put up with it.'

'I am not afraid of a water-sprite,' said the bearward with a chuckle. 'But I have walked a long way today, and I have no mind to walk even a few miles more. I am tired, and so is Braun here.' He jerked a thumb over his shoulder towards the shaggy brown bear on the end of a chain.

'Very well,' said the miller, 'come in, and most welcome you will be. But never say I did not warn you.' He stood aside to let the bearward through. 'You had best bring your bear in with you, you never know what may happen to him if you leave him in the barn.'

The bearward laughed. 'Braun can take care of himself. He would make short work of any water-sprite, I warrant. Just you let us have a sight of this plague of yours and we may be able to do you a service, Braun and I.'

The water-sprite had already had his supper and gone back to the stream, so the two men had a pleasant evening together, chatting of this and that, and since the water-sprite did not appear again before they went to bed, to the miller it seemed quite like old times, and he began to feel more cheerful than he had felt for months.

The miller and his guest shared the big bedroom, with the bear curled up on the floor beside the bed; and there was no sign of the water-sprite all night.

They got up early in the morning and went down to the kitchen for their breakfast, the bear coming after them. But, early as they were, the water-sprite was earlier. There he was, sitting by the hearth, the embers raked together to make a good fire for him to cook his breakfast on, and four broiled fishes laid out on the floor beside him in a row, ready for him to eat when he had cooked the fifth.

The miller's face fell, and the morning suddenly seemed not so bright and pleasant after all. 'There he is,' he whispered miserably.

The water-sprite looked up, showed his teeth and, 'Good morning, miller,' he said. He gave one glance at the stranger, saw nothing to interest him there, and went on with the cooking of his last fish.

The bearward watched him for a moment, then he turned to the miller, winked, and called back over his shoulder, 'Come on, Braun, here is your breakfast for you. Look, good fish.' He pointed, gave the bear a push, and the bear ambled towards the hearth, sniffed at the fishes laid out on the floor, liked what it smelt, picked one up and swallowed it in one bite—one bite more quickly than the water-sprite could have managed. Before the water-sprite realized what was happening, the bear had taken a second helping; but before it could manage a third, the water-sprite had jumped to his feet, quite furious, and was shaking his fists at the bear. 'Away with you! Away with you, you thieving creature!'

The bear sat up and looked at the water-sprite, but made no attempt to go away. Instead, after a moment, it made a move to take the third fish.

'My fish! My fish!' screamed the water-sprite, beside himself with rage. And he rushed at the bear to drive it away.

The bear put out a great paw and clouted the water-sprite, who shrieked and turned tail, making for the mill door and his stream as fast as he could, followed

by the growling bear. At the door the bear turned and came back for the remaining three fishes, which it ate happily by the fire; while the miller, delighted with the way things had turned out, made the breakfast.

The bearward laughed loudly. 'Well,' he said, 'did I or did I not tell you that Braun would make short work of any water-sprite?'

After breakfast the bearward went on his way to the village and the miller began his work, feeling happier than he had felt for a very long time. In fact, he felt so happy that he sang at his work, a thing he had long forgotten to do.

All that day the water-sprite never showed himself near the mill, and it was the same the next day, and the next, for nearly a week; and the miller was feeling on top of the world and thinking of taking a day off from work the very next morning and going over to his wife's mother's house to tell his wife to come home, when, coming whistling into the kitchen for supper, he saw the water-sprite sitting by the hearth cooking his fish.

The miller could have wept. Now it would start all over again, he thought. No wife, no boy to help him, no serving-wench, no peace in the mill ever again.

'Good evening, miller,' said the water-sprite, showing his long teeth. But the miller had not even the heart to give him a civil answer—though he usually did, just in case the water-sprite took offence: one never knew, and it was always best to be on the safe side, with those long teeth.

The miller sat down at the table, too miserable to trouble about getting himself any supper. After a time the water-sprite said, 'That big cat of yours with the long claws, miller; I have not seen it for several days. Has it gone away?'

For a moment the miller went on staring at the table top, thinking regretfully of the bear; not even finding the idea of a bear-sized cat amusing, though it would once have made him laugh. Then suddenly his heart gave a great bound, for he was really quite a quick-witted man. He looked up and said, as casually as he could, 'Why no! She has just had kittens. You will be seeing her around again soon, and all the seven little ones with her. They are just like their mother, only smaller. But they will soon grow. They will soon grow.'

The water-sprite looked at the miller with his round, pebbly eyes even rounder than usual, if that were possible. Then he dropped the fish he was cooking and sprang up. 'Is that so? Seven little ones?' he said with a shriek. 'Then I am off! Good-bye, miller, you will not be seeing me again.' And he was away out of the mill as fast as his flat feet could carry him, and into the water and down the stream and away for good and all.

And the miller never set eyes on him again.

IV

The Seven Proud Sisters

ONCE, in the castle of Schönberg, on the banks of the Rhine, lived a count and his seven daughters: seven maidens as proud as they were beautiful. So proud they were that they considered no man worthy of them—no, not if kings had come a-wooing. They scorned and rejected all their suitors: but not without first tormenting each one with hope, so that they might laugh to see his sufferings.

Seven honest yeomen came to Schönberg, and the eldest sister said, 'Should we demean ourselves by listening to you?' And away the good yeomen were sent, insulted and humiliated and mocked at.

Seven rich merchants came to woo them, and the second sister said, 'Can your riches make you the equals of the daughters of a count?' And away the merchants were sent, downcast and despised and thinking themselves poor for all their wealth.

Seven gay squires came merrily to the castle, young and fresh as the woods in May, and each one of them handsome enough to please any maiden. But the third sister said, 'You are foolish striplings, unproved in battle. Do you think your fine clothes make men of you? Your wooing offends us, begone.' And away the squires rode, their merry songs forgotten, the feathers in their caps drooping like their spirits.

Seven brave knights rode up to Schönberg, eager and full of hope, having valiant deeds to boast of. But the fourth sister said, 'You are mere knights. You dishonour yourselves by your presumption.' And away the brave knights were sent, believing themselves of no account and unworthy.

Seven barons with their men at arms came boldly to the castle, and the fifth sister said, 'Our father is a count, shall we shame ourselves by looking lower for our husbands?' And the barons, rejected and despised, rode away with the gifts that they had brought.

Seven counts made their way to Schönberg with their lordly following of squires and pages, men at arms and knights, and a noble show they made. But the sixth sister said, 'Are we to look no higher than you? Your lands are not wide enough for us. We have no more to say to you.' And away the counts were sent, bewildered and distressed and doubting their own worth.

Seven dukes came to the castle with their train of

courtiers and their chamberlains and their generals and armies stretching miles along the road to Schönberg. But the youngest sister said, 'You are but dukes, though you think yourselves mighty enough. Our beauty is not to be wasted on such as you. Were you princes, we would have none of you. Farewell.' And, hurt and indignant, the seven dukes rode away, disgraced before their courtiers and slighted before the world.

And so the seven proud sisters dwelt in the castle of Schönberg, scorning all wooers, yet never unwooed, such was the fame of their beauty.

One day a young minnesinger, a troubadour, came to Schönberg and sang songs and made verses in honour of the sisters and they let him remain among them, for it amused them to be praised. He sang songs about them all and praised them all, but his eyes turned the oftenest upon the youngest sister, and she marked this well. 'It would be a rare jest,' she said to her sisters, 'to break his heart and spoil his singing for him, this foolish poet. I will let him believe that I return his love—for a little while.'

So the youngest sister smiled on the minnesinger and sometimes she looked long at him, and once she dropped a rose for him to pick up. Deceived by her wiles, he thought that she loved him, and he was happy. And then one day he dared to ask if he might speak to her alone, and she whispered to him a time and a place where he might meet her and say what he would.

She told her six sisters and they hid behind the hangings in her chamber, and when the minnesinger came to her, she dismissed her waiting-women and smiled at him. He thought they were alone and he knelt at her feet and offered her his heart, telling her of his great love as only a poet could tell it. She heard him through to the end and then she began to laugh, and from all around the room her laughter was echoed. Then out from behind the hangings stepped the six elder sisters, jeering and mocking and mimicking him.

'That you,' they said with contempt and scorn, 'you who are nothing more than a poor poet, should dare to love one of us. Away with you, for we are tired of your songs. Away with you for ever.'

The unhappy minnesinger rose from his knees and stumbled from the room. Out from the castle he fled, hearing still their laughter in his ears. But the heart of a poet breaks more easily than the heart of any other man, and he could not live with his grief and sorrow, so he fled to the bank of the Rhine. 'May they one day pay for their cruelty,' he cried, and leapt into the river and was never seen again.

The seven sisters laughed when they heard of the minnesinger's end. 'We wish the Rhine joy of him,' they said. 'For our part, we had grown to find his company tedious and his songs had become stale.'

But the Rhine paid heed to the young minnesinger's prayer and avenged the wrong that had been done to him. One day when the seven sisters went to walk along the bank of the river, they never came home

again. Some say that as they walked, a great wave rose up from the water and swept them away, so that they were drowned, crying out for help which never came. But others say that as they walked, seven princes of the water folk stepped out from the river, and each one took the hand of a sister in his own cold hand and led her, whether she would or no, into the Rhine, on, on, deep, deep down, below the green water.

Yet, however it was, the seven proud sisters were never seen again in their father's castle. But when the Rhine is low near Schönberg, still to this day can be seen seven rocks, each one no harder than the hearts of the seven proud sisters.

V

The Heinzelmännchen

In the city of Cologne, in the old days, lived the Heinzelmännchen, little beings who delighted in kindly deeds, helping servants and busy housewives with their chores and craftsmen at their trades. There are tales of them from many parts of Germany, but it is said that they gathered thickest at Cologne, because for some reason they loved the old city: and certainly the citizens of Cologne profited by their choice.

To almost every house at least one Heinzelmännchen would come by night, unseen and silent, to finish a task that had been left unfinished, to start the chores for the morrow, or to do half or more of his work for someone who was pressed for time.

In Cologne the housewives and the servants had an easy life, and the craftsmen were speedier and neater at their work than any others throughout Germany:

and all because of the willing help of the Heinzelmännchen.

The bakers who had left their dough to rise overnight never needed to leave their beds at dawn in Cologne, for instead of risen dough awaiting them, they would find crisp-baked loaves of white wheaten bread or black rye bread, all ready to take out of the ovens.

The cobblers had only to leave their leather cut upon their benches, and, in the morning, there would be the shoes all neatly sewn with tiny stitches.

Before a holiday, or when guests were expected and the housewife wished to have her home all clean and shining, clean and shining it was, thanks to the Heinzelmännchen: pots scoured bright and tables polished, fresh rushes on the floors and every little pane in the windows as clear as crystal.

Among the townsfolk the Heinzelmännchen had their favourites, and for them they would work doubly hard and even bring them gifts—usually taken from the house of a neighbour less well liked.

When one of their favourites married and brought home his bride, the Heinzelmännchen were not idle. They furnished quite half his house for him with things from other people's homes, so that on his wedding day, the lucky bridegroom could look around his house and count himself fortunate to have so many Heinzelmännchen friends to give him wedding gifts. While in other parts of the city householders would be searching diligently for this or that, wondering how they could possibly have come to lose so large a thing as a stool or a dish; blaming the cat for taking the roasted goose or the neighbour's dog for running off with a side of bacon; and asking themselves if they had really left that beggarman alone at the door long enough for him to have stolen a pair of ladles.

I cannot tell you what the Heinzelmännchen were like, for no one ever saw them. They came after dark and left before dawn, and were quite invisible by day. But, naturally, there was much speculation amongst those they helped as to what their unknown friends could be like; and there was none more curious over the matter than the wife of a certain tailor whom the Heinzelmännchen greatly favoured, so that with their help he had won great prosperity and happiness. Yet it is bad for a man to have an inquisitive wife, as the tailor was to find.

At first the tailor's wife only asked questions of everybody who might have an answer: her neighbours, the priest, strangers passing through the city, and anyone else whom she thought might possibly be able to satisfy her curiosity. But her neighbours only said, 'We know no more than you.' And the priest said, 'They are heathen creatures. One should not meddle with them.' For the priest—good man—was a priest and neither a busy housewife nor a craftsman with a living to earn; and besides, he thought that it was his housekeeper who kept his home so clean and bright, and brushed his cassock ready for him to put on every morning. And the strangers said, 'Why question us?

You live in Cologne, you should know more of the matter than we.'

So then, in spite of her husband's protests, she took to creeping out of bed at nights and going softly downstairs to see if she could catch a glimpse of the Heinzelmännchen at their tasks. But they were always gone before she came. One night she even hid herself in the kitchen and pretended that she had gone to bed, keeping as quiet as a mouse—indeed, far more quiet than a mouse—hidden behind a settle in the dark. But the Heinzelmännchen were not deceived and they never came that night, nor on any other night when she waited up for them. And each time dawn found her, cold and cramped and cross, her curiosity even stronger than before.

At last, one morning, she had an idea, and passed all that day waiting impatiently for the night. She never told anyone of her plan, which was a pity, for if she had, someone might have warned her not to try it. Her husband most certainly would have done so, for he was well pleased with the help of the Heinzelmännchen, and would never have risked offending them.

That night, as soon as all the household was abed, the tailor's wife slipped down to the kitchen and took a jar of dried peas from the shelf, then she went upstairs again carrying it and strewing as she went a handful of peas on every stair. At the top, the jar was empty. She returned to her husband's bed, delighted with her cleverness. 'If that does not do the trick, then nothing will,' she said to herself. 'One of them is bound to step on a pea and fall downstairs and hurt himself, and there he will be in the morning, unable to get away when the others go.' And she dropped off to sleep beside her snoring husband and had pleasant dreams.

At the first light of dawn the tailor's wife jumped out of bed and never even stopped to put on her shoes, she was so eager. She shuffled across the rushes, ignoring the tailor's muttered, 'Surely it's not time to get up?' and picked her way carefully—and painfully—down the pea-strewn stairs. But the Heinzelmännchen had not fallen into the trap. They had come that night—certainly they had come—but they had gone away again, leaving the work undone, disgusted at such discourtesy. And what is more, they never came to that house again.

Bitterly the tailor's wife regretted her foolish curiosity, as day after day she had to scrub and sweep and polish, and bitterly she regretted it each time her overworked husband nagged at her and said, 'It was all your fault'—which he did a dozen times a day.

But the tailor's wife was not the only one whose curiosity over the Heinzelmännchen got the better of her common sense and her gratitude, and in time there were fewer and fewer houses in Cologne where the Heinzelmännchen still went; and after many years there came a day when they left the city altogether. They went in broad daylight, all of them together, marching through the streets to the sound of drums and pipes. People heard the music and said, It is the Heinzelmännchen going,' and they ran to doors and windows to see them go, while folk in the streets stood and stared. But even on that last day the Heinzelmännchen kept their secret. For though everyone who was not deaf could hear the music and follow its sound along the streets to the city gates, there was not a single Heinzelmännchen in sight.

The music passed through the gates and was lost in the country beyond the city, and no one knew where the Heinzelmännchen had gone. But wherever they went, they never came back to Cologne, and the city has never been so happy or so lucky since.

VI

The Ratcatcher of Hamelin

A LONG time ago—in the year 1284—it is said, the town of Hamelin, in Brunswick, was plagued with a great host of rats and mice. They ate up the corn in the granaries, they ate up the food in the houses; and however many were caught by the townsfolk and by the cats and dogs, there always seemed to be more to take their place; so that very soon everyone in Hamelin was in despair. 'Soon we shall be starving,' they said, 'for what will happen when the rats and the mice have eaten all the food?'

The burgomaster and the town councillors worried day and night. All day they never had a second's peace from the folk who came, one after another, to ask their help and advice. And at nights they could not sleep—even in their comfortable beds—for the thought that at that very moment the mice would be in their own storecupboards and the rats would be eating the oats in the mangers in their own stables and the bacon hanging from their own kitchen beams.

One morning in the month of June a stranger came to the town hall and asked to speak with the burgomaster and the councillors. He was clad most curiously, in a tunic and jerkin made of cloth of every colour, the like of which they had never seen before. 'I am a ratcatcher,' he said. 'If you will pay the price I ask, I will rid your town of rats and mice.'

'If you can truly do as you say, then no price will be too high for us to pay,' said the burgomaster.

'Name your price, stranger,' said the councillors.

So the ratcatcher named his price, and without any ado the burgomaster and the councillors agreed to it. 'When will you set to work?' they asked eagerly.

'Now,' he replied.

He went out into the market place and took from his pouch a little pipe. On this pipe he began to play a tune. Soon, out from every house in Hamelin, came the rats and the mice, till the cobbles of the market place were dark with them. Then the ratcatcher, still playing his pipe, moved along the streets, and the great crowd of rats and mice followed him.

Away went the ratcatcher, piping, down to the banks of the River Weser. Here he kilted up his many-coloured tunic and waded out into the river, with all the rats and mice after him. And in the river every single one of them was drowned. Then the stranger returned to the town hall for his payment.

But, now that the rats and the mice were gone, the burgomaster and the councillors were less willing to pay the price they had agreed to pay, and they began to excuse themselves to him on one pretext or another.

'It would leave our public coffers almost empty, if we paid you what you ask,' said one.

'Surely,' said another, 'you cannot demand so high a price for a task that has taken you so short a time to perform?'

'Had the Weser not been so near and handy, what would you have done?' asked yet another. 'The river did more than half your work. No more than half payment is due to you.'

And so they went on, while the ratcatcher grew more and more angry. 'Very well,' he said at last, 'keep your gold. I will take my own payment, and at my own time,' and he went from the town hall and away out of the town.

The burgomaster and the councillors thought that they were rid of him, as they were rid of the rats and the mice; and they were well pleased with their bargain.

But the very next Sunday, while all the folk of Hamelin were in church, the ratcatcher in his many-coloured garments and wearing a strange red hat, came again to the town. In the market place he stood and played his pipe, and out from every house there came, not rats and mice this time, but all the children of Hamelin, left at home while their parents were at mass. They gathered about the ratcatcher, laughing and singing and dancing; and when he thought that they were all there and none were left in the houses, he walked along a street away from the market place, still playing on his pipe; and all the children followed him.

On to the town gates he went, out through the gates and away across the fields beyond, on towards the hill called the Koppelberg. And when he reached the Koppelberg, the hill opened to him and he went in, and after him went all the children; and the hill closed on them and they were never seen again.

Of all the children in Hamelin, only two were left, because they had lain abed too long on the Sunday morning and had stopped to dress themselves when they heard the pipe. They ran after the others, but reached the Koppelberg too late, when it had closed again; and so they alone returned safely to their homes.

One hundred and thirty children were lost that day, and among them the young daughter of the burgomaster, who now bitterly regretted the fine bargain he and the councillors had made.

VII

Till Eulenspiegel

Here, now, is the tale of a rogue with a merry heart, a liar, a cheat and a trickster, who yet was loved by all who enjoyed a jest and thought laughter the cure for most ills.

Long ago there lived in a village in Brunswick a poor peasant named Klaus Eulenspiegel, with his wife, Anna. To these good people, one day, was born a son; at which they rejoiced, for they had no other children. When the time came, Klaus and Anna, together with their friends and neighbours, took the child to church to be christened, and he was named Till. On the way home again, the woman who was carrying the child slipped and fell, dropping Till into a muddy puddle, so that when they reached Klaus's cottage, Anna had to fill a basin with warm water and wash Till from top to toe.

The neighbours all laughed and said, 'He will surely grow up into a great and famous man, for has he not been christened three times in one day? Once with holy water, once with puddle water, and once with warm water out of a kettle. Surely only the most remarkable people are christened more than once?' And Till did indeed grow up to be famous—but not in quite the way these good folk hoped.

From his earliest years Till was a merry child, always laughing, always playing and always teasing the other children in the village. Until, before very long, all the villagers were agreed that there was no boy in the district more full of fun and naughty ways than Klaus Eulenspiegel's son.

But though Till might be quick to learn a new trick or a jest, he was quick to learn nothing else, and paid little enough heed when his father would have taught him how to plough or sow or reap; and he was rarely by when his mother wanted help with some household task: drawing water from the well or fetching sticks for the fire.

One sad day, when Till was some sixteen years old, Klaus died, leaving his widow in great poverty. Anna wept, exclaiming, 'What will become of us? If only you had been a dutiful, diligent son, and learnt a trade, you would be able to earn enough to keep us now.' But Till, not one whit abashed, cared nothing for her words.

There came a morning when the last crust was eaten and the last coin spent, and Till's mother said to him, 'Unless we have bread we shall starve. Alas, that I should have lived to see this day!' And she hid her face in her apron and wept.

'Bread?' said Till. 'Is that all you want? Dry your tears, mother. You shall have bread.'

He went to the nearest town and found out the best baker there. Setting his cap at a jaunty angle and straightening his threadbare jerkin, he swaggered boldly into the baker's shop. 'Good morning to you, baker. My master, the great count, is passing through your town. He is putting up at the best inn, and needs bread for himself and his servants for the next stage of his journey. A crown's worth of loaves is what he needs. Give me your boy to help me carry the bread to the inn, and my master will pay him the crown.'

It was not every day that the baker sold a crown's worth of bread to a single customer. He bowed low to Till, called his boy and hastened to fill a sack with his best white loaves.

Till and the baker's boy set off, carrying the bulging sack between them. Half-way to the inn, as if by accident, Till contrived to spill one of the loaves into the roadway, so that it fell in the mud. 'Here's a fine to-do,' he said. 'My master is very particular, he will be in a rare temper if he sees this loaf. Run back to the shop and fetch another in its place, and I will wait here for you.'

The boy ran off, and immediately he was out of sight, Till slung the sack over his shoulder and made for home as fast as he could. 'Here is bread enough to last us for a fortnight, mother, and some to sell besides. Good white bread, too. What do you think of me for a son, now?'

But even the best of tricksters dares not play the same trick more than once, so when all the bread was gone and Till's mother began to weep again and deplore his laziness once more, Till said, 'Dry your tears, mother. I shall go to Brunswick city and find work.'

So Till set off for Brunswick city and there he hired himself to a baker, saying that he was a skilled baker's man.

For two days Till helped the baker in his shop, and on the afternoon of the third day, the baker said to him, 'I have to be away from home tonight. I cannot return until morning, so you must do the baking for tomorrow.'

'I will do the baking, master,' said Till. 'But what shall I bake?'

The baker, who had a quick temper, lost it. 'What will you bake? What do you think you will bake? Owls and monkeys? And you a skilled baker's man! Let me have no more of your nonsense.' And away the baker went.

Till at once went to the bakehouse and set to work, and all the dough for the next day's bread he fashioned into the shape of owls and monkeys and put them into the oven to bake. When the baker came home again, early the next day, he found not a single loaf baked, but only owls and monkeys. 'Fool of a lad,' he stormed, 'what have you done?'

Till opened his eyes very wide and looked as innocent as he might. 'Why, only what you bade me, master. When I asked you, you bade me bake owls and monkeys, and so I have done.'

The baker boxed his ears. 'Away with you, and never let me set eyes on you again,' he roared. 'But first pay me for the dough that you have wasted, or I shall have the law on you.'

'If I pay you for the dough,' asked Till, 'will the owls and the monkeys be mine?'

'You can do what you please with them,' said the baker. 'I want none of them.'

Till had his two days' wages and a coin in his purse besides, so he paid for the dough, filled a sack with the owls and the monkeys and went, sped on his way by a clout from the still angry baker.

Near the big church Till sat down beside his sack, wondering how best he might get rid of his owls and his monkeys and make some money in the doing of it.

Now, it was the eve of St Nicholas' day and the townsfolk were going into church. Till grinned to himself and went and stood by the church doors. As the people came out, he showed them his wares. 'Will you buy an owl or a monkey for the feasting tomorrow, good people? Will you buy an owl or a monkey?'

Everyone laughed at the quaint fairings he was selling, but most of them stopped to buy, and in a very short while his sack was empty and his purse was full, and he and his mother were able to celebrate St Nicholas' day together in fine style.

Till never did a stroke of work if he could help it, but his ways of making money from unsuspecting folk were many and ingenious; and before long he had enough to buy himself a horse, and he rode all about Germany playing his tricks.

One day, at Bremen, he went to the market at the time when the countrywomen were carrying in their pails of milk to sell in the town. With a big barrel beside him, Till stood in the market place calling out that he would pay a higher price for milk than anyone else. At his words all the women came to him, and he told them to pour their milk into his barrel. 'When my barrel is full,' he said, 'I will pay you.'

Pleased to be selling their milk with so little trouble, one after another, the women did as he said, then sat down close by, chatting and gossiping until such time as the barrel should be full and the paying should begin.

But when the barrel was full to overflowing and could not hold a single drop more, Till said, 'That is all that I can buy from you today, goodwives. Each one of you shall have a fair price for her milk, as I promised. But I have no money in my purse today, so you must all give me credit until this day two weeks.'

The women cried out at this, and Till shouted above their noise, 'Anyone who does not wish to trust me, let her take her milk away again.'

At this all the women rushed for the barrel, each one

trying to be the first to refill her pail; so that what with the pushing and squabbling, hair being pulled and sharp elbows being dug into ribs, the milk was all spilled in the market place. In the confusion Till made his escape, laughing fit to split his sides; and the countrywomen had to go home with a sad tale to tell their angry husbands.

One day Till found himself with no money and nothing to sell but his horse. He had no wish to sell his horse, so he set himself to devise a means by which such a misfortune might be avoided. Suddenly he grinned, mounted his horse and rode off to an inn. He stabled his horse in the inn stable and went out to the market place, and there he stood up on a horseblock and called out to all the people that he had a marvel which he would show to them. 'I have a horse,' he said, 'like no other horse in all Christendom. Its tail is where its head should be; and where its tail should be, there is its head. For one small coin—one small coin only, good people—I will show this marvel to you.'

Few folk can resist a marvel, so they crowded round him, dropping their coins into his cap which he held

out. When the cap was full, Till led the way to the inn. 'Here is my horse,' he said. 'Tell me if you do not think that its head is where its tail should be and its tail in the place of its head.' And he flung open the stable door, and there inside was his horse—its tail tied to the manger and its head looking out through the door.

Luckily for Till, there were many there who saw the joke and laughed heartily at it, for all the coins that they had lost; and their good-natured laughter won over those who would otherwise have been angry at being fooled.

Here is a trick which Till played at Hanover, when he had one day ridden there. By the city gates he came upon twelve blind beggars, asking alms. 'Good friends,' said Till, 'would you eat and drink your fill and lodge for a few days at the best inn in the town?'

'We would indeed,' replied the beggars. 'But no such luck will ever come our way.'

'And why not?' asked Till. 'Charity is a virtue, very pleasing to heaven. I will be charitable to you, my friends. Here are twelve crowns, one for each of you. Go now to the best inn in the town and spend them.'

Because the beggars were blind, they could not see that Till did not hand a single crown to any of them, but each of them believed that one of the others had the money. They would have stayed to thank him for the rest of that day, had he not hurried them off to the inn.

When the surly innkeeper saw them coming, he

called out, 'I want no beggars here. Be off with you.'

But when one of them spoke for them all and said, 'We are not begging, good master host. We can pay for all you have to offer. A rich gentleman has given us twelve crowns to spend in your inn,' then the innkeeper changed his tune.

'Twelve crowns!' he exclaimed.

'Yes, indeed, twelve crowns,' said the blind men. 'Such generosity has never come our way before.'

'And it is never likely to come your way again,' thought the innkeeper to himself. 'That gentleman must have been touched in his wits.' Out aloud he said, 'You are most welcome, sirs. Come you into my humble inn, and the best I have shall be yours until the twelve crowns are spent.' And he showed them into his finest room with many bows, which were quite wasted on them, since they could not see.

For almost a week the blind men ate and drank of the best, and lay on a soft bed at nights, until the innkeeper, greedy for payment, decided that they had had their twelve crowns' worth. 'It is time for you to go, sirs,' he said. 'Now pay me my reckoning. Twelve crowns you owe me.'

'All good things must come to an end, I suppose,' sighed each blind man. 'Blessings on the fine gentleman who gave us this good time.' But not one of them made a move to pay the innkeeper.

He became impatient. 'Come, now, my money, please.'

'Who has the money?' asked one of the blind men.

'Not I,' said another.

'Nor I,' said a third.

And so they all said, all twelve of them; and the innkeeper realized that their fine gentleman had played a trick on him.

In vain the blind men pleaded that it was none of their doing and that they had not meant to cheat him. 'That is nothing to me,' he said. 'Here you stay until the reckoning is paid, or to prison you all go. But you leave my best room this instant.' And down the stairs and out to the yard he chased the blind men and thrust them into the pig-sty and bolted the gate on them.

Meanwhile, Till was wondering how the blind men had fared. 'I had best go to find out,' he thought. 'The innkeeper may not be a man who laughs easily.' So off he went to the inn and entered as though he would have asked for lodging, and there in the yard he saw the blind men, shut in the pig-sty, very cold and miserable and hungry; and he was indignant. 'A fine place to lodge your guests!' he said.

'They are no guests,' said the innkeeper. 'They are rogues who cheated me.' And he told his tale to Till.

'If someone might be found,' asked Till, 'who would discharge their debt for them, would you let the twelve blind men go free?'

'Indeed I would, for it is my twelve crowns that I want, not twelve blind beggars. They can go hang themselves for all I care.'

'I will find someone who will go surety for them,'

said Till. 'Out of all Hanover, that cannot be impossible.'

Till went to the house of the parish priest and said to him, 'Good father, at the inn where I am staying the innkeeper is in great distress, for he is possessed by a devil and has sent me to ask your help. Can you cast the devil out of him, and give your blessing to this poor man?'

'Indeed, I will do all he asks, my son,' said the priest. 'Tell him to come to me as soon as he may.'

Till returned to the inn and said to the innkeeper, 'I have found a good priest who will pay you what the blind men owe you. If you doubt my word, send your wife to him to ask him if I lie.'

Mightily pleased, the innkeeper sent his wife with Till to the house of the priest, and there Till said, 'Here, good father, is the wife of that unfortunate innkeeper I told you of. I beg you, assure her that you will do as he asks.'

'Why, certainly, good woman, I will do as your husband asks, and speedily. For it would be a most charitable deed, to set his mind at rest, the poor afflicted man.'

The innkeeper's wife bobbed her thanks and hurried back to her husband. 'The stranger spoke truly,' she said. 'All will be well, for the priest will pay the debt.'

'Now let the blind men go,' said Till.

The innkeeper opened the gate of the pig-sty and hustled them away—and very thankful they were to be free—so that he might go to the priest at once and collect his twelve crowns.

But when he presented himself at the priest's house, the priest made ready to cast a devil out of him, not to give him twelve crowns; and when he said that he had come for his money, the priest said, 'Yes, yes, that is always the way of devils, always demanding the riches of this world. Be calm now, and with God's help you will soon be in your right mind.'

'I want my twelve crowns which you promised me,' said the innkeeper; but the priest paid no attention. Then the innkeeper lost his temper and began to threaten the priest, who was thereby even more convinced that he was possessed by a devil. And so it went on with the two of them at cross purposes, until the innkeeper saw that he had been tricked and, knowing that he would never see his twelve crowns, he went home in the worst of tempers, grumbling curses on Till and the twelve blind men. But little good did that do him, and it certainly never brought him twelve crowns.

There was another innkeeper, in Eisleben, whom Till served a fine turn. This innkeeper was always boasting and setting himself up above his fellow men, and one wintertime, when the snow lay thick, as Till passed through Eisleben, he lodged at this man's inn. That evening, after dark, three merchants, on their way to Nuremberg for the fair, came late to the inn. They looked pale and shaken, and cast many glances backward over their shoulders even in the safety of the town, and when Till asked them what ailed them, they replied, 'Along the road, in the darkness, a great wolf sprang on us, and much trouble we had to drive the beast off.'

The innkeeper burst out laughing. 'One wolf, and three of you, and you were afraid. If I, alone, met two wolves, I would fight them off with my bare hands, or maybe kill them. To make so much pother over one wolf! But perhaps in the town where you come from, all the men are cowards?' And so he mocked the three of them all the evening until they went to bed.

Till shared a room with the merchants and he said to them, 'Would you see that boastful knave taken down?'

'We would indeed.'

'Then,' said Till, 'lodge at this inn on your way back from Nuremberg. I will be here too, and we shall see what we shall see.'

In the morning, the three merchants went on their way, and Till went on his.

While the merchants were selling their goods at Nuremberg, Till went hunting and killed a wolf, then buried its body in the snow until he should need it. At the time when he thought the merchants would be returning from Nuremberg and passing again through Eisleben on their way home, Till dug up the dead wolf, put it in a sack, and carried the sack to the inn. There, as he had hoped, were the three merchants, and the innkeeper was once again mocking them and asking them if they had not yet been eaten by wolves. 'You should see me with a wolf,' he said. 'That would show you how a brave man deals with such beasts.'

Till greeted the merchants, gave them a nod and a wink when the innkeeper was not looking, and they settled down to sup together. After supper they went to bed, and Till told them his plan.

When all the household was fast asleep, Till took his sack down to the kitchen and there he opened it and took out the dead wolf and set it up by the fire, propped up with a broomstick; and in its mouth he put a shoe. Then he went back to bed.

After a little time, one of the merchants, on the word from Till, went to the door of the room, opened it and called out, 'Good master host, we are thirsty. Tell your man to bring ale to our room.'

His shouts awoke the innkeeper. Grumbling to his wife, he got out of bed and called to his maidservant to fetch the guests some ale. Then he got back into bed.

The maidservant got up, put on her clothes and went

down to the kitchen, yawning and sleepy. When she saw the wolf in the firelight, she gave one little scream and ran out into the snowy yard.

The merchants and Till all cried out, 'Where is our ale? Hurry, hurry with our ale, for we are thirsty.'

'The wench must have fallen asleep again,' thought the innkeeper, and he called to his man to fetch ale for the guests.

The manservant put on his clothes and went down to the kitchen, and when he saw the wolf in the firelight, with the shoe in its mouth, he thought, 'Heaven help us! It is a wolf, and it has eaten the girl.' And he took to his heels and ran for the cellar.

When Till and the merchants still went on demanding their ale, the innkeeper thought that the man, too, must have fallen asleep again. Grumbling more than ever, he got out of bed, lit a candle and went down to the kitchen himself. When he saw the wolf, his hands shook so much that he dropped the candle. Without another glance, he fled upstairs calling to Till and the merchants, 'Come quickly and save me, my good friends, I beg of you. In the kitchen stands a great monster wolf and it has eaten the man and the maid. If you do not kill it for me, it will eat me also.'

Then Till and the merchants went down to the kitchen and showed the innkeeper that the wolf was dead. 'If you run from a dead wolf, my brave one,' said Till, 'what would you do if you met with a live wolf, like my three friends here?'

Then the innkeeper's wife came down from her bed and the man crept up from the cellar and the maid-servant in from the yard, and they all laughed until the tears came at Till's good jest—all save the innkeeper himself. He went quietly back to bed, and it was many a long day before he boasted again.

But, alas, it is not everyone who can enjoy a laugh, and there are some unfortunate people who prize dignity above a sense of humour; so there came a day when Till found himself in prison in Lübeck, charged with stealing wine from a tapster by a cunning trick. He was brought to trial, and though the townsfolk themselves would have been ready enough to pardon him, the city councillors, who thought too much of their dignity, declared that, for the good name of the city, Till should be hanged.

When the time came for the hanging, a huge crowd had collected before the gallows, and so great was the fame of Till's tricks and jests, that there were many there who thought that he might yet save himself in some way; and, indeed, they hoped that he might.

But when Till was brought from the prison, he appeared meek and subdued and as though he had lost all hope. When the hangman put the rope about his neck, he asked humbly that he might speak. 'In a little time,' he said, 'I shall be dead, and justly, for my crimes are many and I deserve to die. Yet I beg of you, good people of Lübeck, to grant me one boon before I die. I do not ask my life of you, nor money, nor anything which it will cost you anything to give. Indeed, by my

request I shall be saving you a small expense. Will you not give a dying man his last wish?'

The city councillors put their heads together, and after long discussion of the matter they decided that as they were being offered a chance of saving public money, there could be no harm in agreeing. 'So long as it is as you say, and you do not ask of us your life, or money; and so long as you do indeed save us an expense, then we will grant your request.'

'I am much beholden to you, good gentlemen,' said Till. 'All I ask is this: that I may make amends for my theft from a citizen of Lübeck by saving the city itself the cost of my death. I see that the rope around my neck is a new one—for that you will be charged a crown or two. And there is the hangman's fee as well—I I would save you that if I could. Give me an old rope, good gentlemen, and let me go quietly away and hang myself, and the public coffers will be the richer by the cost of a rope and the hangman's price.'

Everyone save the hangman and the ropemaker thought this a sensible suggestion, and before many minutes had passed the new rope was taken from around Till's neck, an old rope was put into his hands, and he came down from the scaffold. 'You have my thanks, good citizens of Lübeck,' he said. 'When I have time enough to spare and nothing better to do with it, I shall hang myself.' And quickly he slipped among the crowd and was safely gone. And there were many who were glad of it, and laughed.

In time Till's fame spread not only throughout all Germany, but to other lands as well; and it came to the ears of the king of Poland, who loved jests and tricks above all things. He had two fools of his own, who were considered unequalled in the world, and one day he sent to Till asking him to come to his court and match himself with them: the prize for the winner to be a fine new coat and twenty pieces of gold.

So Till went into Poland and presented himself before the king, in company with the two Polish jesters, who had no doubt at all that they could easily surpass him.

'Now,' said the king, 'to that one of you who can wish the greatest wish, I shall give the prize. Come, what are your wishes?'

'This,' said the first Polish jester, 'I wish that the sky were paper, and the sea ink, and that on that paper with that ink I could write down how much money I would like for myself, and that I should have it.'

'That is good,' said the king. 'Very good.' And he laughed.

'And I,' said the second Polish jester, 'wish that I had as many castles as there are stars in the sky, that in them I might keep the money which my friend here has wished for.'

'That is even better,' said the king, and he laughed again. 'Surely there cannot be a greater wish than that?'

'I wish,' said Till, 'that the two of you, my friends, might make me your heirs, and, on the very same day, that the king might hang you both.'

And at that the king laughed until he could laugh no longer, and he gave the prize to Till, who was well enough pleased with his new coat and the money.

And so Till went on all through his life, with his tricks and his jests; but at last there came a day when he was very old and knew that he must die. Lying on his bed in his home in the town of Mölln, he sent to the parish priest and to certain of the townsfolk, saying he wished to make his will. 'A third of my wealth shall I leave to the church, a third to my friends, and a third to the town of Mölln,' he said to them. 'My wealth lies in that coffer there'—and he pointed to a large box beside his bed—'it will be yours when I am gone. But one last request I would ask of you: do not open the coffer until I have been buried a month.' And this they all promised him.

A short while after, Till died, regretted by many. He was buried with all honour in the town of Mölln, and above him was set a fine gravestone carved with a likeness of himself, holding in his hands an owl and a mirror, in remembrance of his name: Eulenspiegel—Owlglass.

But those who were to be his heirs, once they had done mourning him, could hardly wait until the month was up before opening the coffer. Day and night they thought of it, wondering and calculating how much would be their share of Till's riches, and planning how to spend the money when they had it.

When the four long weeks were over at last, they gathered in Till's house and unlocked the heavy coffer; each one of them, parish priest, friends and town councillors, trying to look less eager than his neighbour. But when the lid of the coffer was raised, inside they found nothing but stones.

Till had played his last trick on them.

VIII

Richmuth of Cologne

In Cologne, in the middle ages, there lived a rich burgomaster whose much-loved wife was named Richmuth. One sad day Richmuth was seized by a sudden sickness, so that she fell to the ground and lay still and cold, as though she were dead. The burgomaster sent for the best doctors in the city, but all to no avail, for none of them could help her, and each one shook his head and said, 'There is nothing I can do. Your wife is dead.'

Richmuth was laid in her coffin; and, as her heart-broken husband took his last farewell of her, he put on her cold finger a favourite ring of hers which had been his gift to her and which she had often worn. When the members of his household sought to prevent him, saying that the ring should surely be kept, along with her other jewellery, for her children, he answered, 'It was her favourite jewel. I did not lend it to her to wear for a few years. I gave it to her for ever. Let her take it with her.' And he would not listen to their protests.

The coffin was carried to the church and Richmuth was buried, and sadly the burgomaster returned home alone. But before the lid of the coffin had been closed, the sexton had noticed the ring on Richmuth's finger and he had thought to himself, 'It is a shame that such a costly ring should be left hidden below the earth. It would fetch a good price, were it sold.' His regrets quickly turned to longing, and his longing to a determination to possess the ring. All day he thought of nothing else, and at midnight he stole silently from his house, went to the church, opened the grave and prised off the lid of the coffin.

There, on Richmuth's finger, was the ring, glinting in the light of the lantern he had brought with him. Covetously he stretched out his hand to take it. But even as he did so, he saw her hand move and grope and take hold of the side of the coffin, as though the dead woman would raise herself up. Terrified, the sexton dropped his lantern, so that it went out, and fled.

Now, Richmuth had not really been dead, but only in a deep trance, and at the moment when the sexton had opened the coffin, she had come to herself again. She sat up in the darkness, wondering where she was and what had befallen her. When she felt the winding sheet about her and saw the tall shape of the church towering above her, she guessed what had happened.

Quickly she unwound the grave clothes and stood up. 'My poor husband,' she thought, 'how unhappy he will be, believing me dead. And my children will be needing me.'

Stumbling, for she was very weak from her illness, she made her way to the street, her one thought to reach her home as quickly as she might. The streets were dark and deserted, and there was no one about who might have helped her. Shivering with the cold, she hurried past the sleeping houses, and at last, after what seemed to her a lifetime's journey, exhausted and well nigh as cold as though she had really been a corpse, she reached her home, and found the big house door barred against her.

She beat on the door with her fists and called till she was hoarse; and at last, by luck, one of the serving-men heard her, came to the door, opened it an inch or two and looked out. 'Who are you, disturbing a house of mourning with such unseemly noise, and at this hour of the night?' he asked.

'It is I, Johannes, your mistress. For the love of God, let me in.'

He held up the lantern he had lighted and peered at her. Her face was white and her hair dishevelled, but it was his mistress all right. And, what is more, she was alive and no ghost. Yet, he thought, how could she be alive, for had he not seen her buried that morning, and wept for her, too?

'I can see it is you, mistress,' he said. 'Yet how can it be you? For you were buried this morning.'

'It is I, it is I, Johannes. Let me in,' she pleaded.

But he was far from being the most quick-witted of the burgomaster's servants, and he just stood there, holding the door open barely a crack, wrinkling his brows and trying to puzzle it out. And it was more than his wits could manage to decide what he should do. 'I shall go and ask the master,' he said to himself. 'He will know what I ought to do.' So he pushed the door to and hurried off up the stairs to the burgomaster's bedchamber.

Poor Richmuth, whose only comfort all through the dark, cold and endless-seeming walk through the city, had been the thought of the welcome she would receive when she reached home, and the great happiness of her husband and her children and all the household, could bear no more. She could not even make the effort to push open the door and walk in. She sank to the ground beside the doorpost, covered her face with her hands and wept.

In his bedchamber the burgomaster lay unsleeping, too sunk in grief to have heard the knocking on the door. 'What do you want, Johannes?' he asked, when the servant came in.

'Are you awake, master?'

'What do you want?'

'The mistress is at the door, master. She is asking to come in. What am I to do?'

The burgomaster sat up in bed. 'You are dreaming, man. Or else you have gone out of your mind.'

'I am wide awake, master; the mistress woke me

with her knocking. And I am in my right mind, but very puzzled as to what I should do. Am I to let her in?'

The burgomaster was too unhappy even to be angry. 'If this is some cruel jest,' he said, 'then may God forgive those who are playing it on me.'

'It is no jest, master. The mistress is standing outside the door. Am I to let her in?'

The burgomaster lay back upon the pillows. 'I would as soon believe that my horses would come out of the stable and into the house through the door, and walk up the stairs to this room, as that my Richmuth were standing outside.' He turned his head away and wept.

The servant stood in the middle of the room, holding his lantern and scratching his head and wondering. The master had failed him. He had not told him what to do. And if the master had not told him, who was there left to ask?

Then suddenly he heard the clattering of hoofs from the hall below and then a trampling on the stairway, and he ran from the room with his lantern to see. There, coming up the stairs side by side, were his master's two white horses.

He ran back to the bedchamber, barely able to speak for excitement. 'Master, master, the horses are coming up the stairs!'

The burgomaster raised his head. 'You are mad,' he said. But then he could hear it, too, hoofs clattering on the wooden stairway and then a loud triumphant neighing as they reached the top. He sprang out of bed, never stopped to pull on his boots or to put on a bedgown, but snatched the lantern from the servant and ran out of the room, giving no more than a glance at the horses outside his door, and was away down the stairs and across the hall.

The house door stood wide open where the horses had passed through, and at first it seemed as though there were no one outside. And then he saw her, huddled on the ground beside the doorpost, weeping. 'Richmuth,' he said, 'Richmuth.' And a moment later she was safe in his arms.

They lived together in joy and happiness for many years more; and, in memory of that night, the burgomaster had two carved horses' heads set upon his house wall.

IX

The Werewolf

THERE once lived in a village two honest peasants who had been close neighbours and good friends for all their lives. We may as well call them Hans and Kunz, since those were quite likely to have been their names.

One day Hans and Kunz decided to go to the forest to fetch wood for their fires. They set out early, each with an empty sack for kindling slung over his shoulder and a length of rope to tie up the faggots. Outside the village they met with a stranger who stopped them with a civil greeting. 'Good day to you, my friends.'

'And good day to you,' said Hans and Kunz.

The stranger looked at the sacks and the ropes they carried. 'I see that you are going to the forest for firewood. I am bound for the forest myself for the very same purpose. May I go with you?' And he showed them a sack and coil of rope which he had with him.

Now, as they told each other afterwards when they talked the matter over, neither Hans nor Kunz cared much for the looks of the stranger, though neither of them could have said why. It was not that his teeth were too long, not that his voice was too deep, nor even that he watched them with a dark, piercing stare. No, it was none of these things: there was simply something odd about the stranger that neither of them liked. But they were both of them kindly, good-natured fellows, never given to blaming a man for the colour of his hair, so they both said, 'Indeed you may. We shall be glad of your company.'

So the three of them went on to the forest; and on the way they talked of this and that and the other, and the road seemed short enough.

In the forest they all three worked hard, and by midday their sacks were full and their faggots tied neatly together. They shouldered their loads and set off for home.

On the edge of the forest lay a meadow in which grazed several mares with their foals. As they came close to this meadow, Hans, trudging along, said, 'Hot work, this.'

'I could do with a rest,' said Kunz.

The stranger glanced at the meadow without appearing to do so and said, 'What say you to sitting down here under this tree and having a rest?'

'I would say it was a good idea,' said Hans.

'I too,' said Kunz.

So they all three laid down their firewood and sat

beneath the tree. The stranger yawned widely and looked at the others. 'I for a snooze. How about you two?'

'Another good idea,' said Hans.

'Well said,' agreed Kunz.

So the three of them lay down and closed their eyes. Within five minutes Hans was snoring peacefully; but Kunz was still awake when he heard a faint rustling in the dry leaves beside him. Lazily he opened his eyes a tiny crack and looked through. He saw the stranger sit up cautiously. 'Aha, what is he up to?' wondered Kunz.

The stranger looked carefully at his two companions, as if to make sure that they were sleeping. He seemed satisfied by what he saw when he looked at Hans, but seemed less certain of Kunz. So, just to reassure him, Kunz gave a snore. This seemed to be all the stranger wanted, for he got up, pulled off his tunic and instantly became a big grey wolf.

Kunz was very startled, but he bravely went on snoring, and the wolf ran off quickly into the meadow. Kunz opened his eyes wide, sat up and looked after him. In the meadow the wolf sprang upon one of the foals and ate it up immediately, hide, hoofs and all. Then he licked his lips with a long red tongue and loped back to the tree.

Kunz hastily lay down and half closed his eyes. The wolf thrust his head into the tunic and instantly became a man again. Then the stranger lay down and closed his eyes and went to sleep, and was soon snoring as loudly as Hans. But, as may be expected, Kunz had never felt less like sleep in all his life.

After an hour or so, Hans and the stranger awoke. 'We had best be setting off for home, I suppose,' said Hans, stretching himself. 'I had a good sleep. How about you two?'

'I had a good sleep, too,' said the stranger.

'I had a strange dream,' said Kunz. But he never told what it was.

They shouldered their loads once again and set off. On the way, as in the morning, they talked of this and that and the other; but not at so great length as before, since now they needed much of their breath for carrying their firewood. And besides, Kunz, for one, had a lot to think about.

As they neared the village, the stranger, who had been walking more and more slowly for the last mile or so, suddenly stopped, laid down his sack and his faggots, doubled himself up, put his hands on his belly and groaned. 'Ah! I have such a pain inside me,' he said.

Hans shook his head sympathetically. 'I am sorry to hear that.'

But Kunz, who could not resist the temptation, leant quite close to the stranger and whispered in his ear, 'I would have a bellyache, too, if I had a horse inside me, hide, hoofs and all.'

The stranger turned and looked at Kunz, a long, dark, piercing stare, then he said slowly, in that very deep voice and showing his long teeth, 'If you had said that to me under the tree, it would have been the last thing you ever said.'

'I know,' said Kunz. He took a tighter hold of his firewood and turned to Hans. 'Let's be going, Hans,' he said.

They hurried off towards the village together, leaving the stranger in the road; and that was the last they saw of him. But you may be sure that they never again went gathering wood with any strangers.

X

The Knight of Staufenberg

A KNIGHT once lived at Staufenberg, near the Rhine. He was young and brave and handsome, even if none too rich. One day, as he rode to hunt in the forest, he lost his way. Searching for a familiar path, he came upon a small, clear spring which gushed out from between mossy rocks. Being thirsty, he dismounted and drank from the spring. When he raised his eyes again from the water, he saw, where there had been no one a moment before, the fairest of all maidens sitting close by, braiding her wet hair.

'Lady,' he asked in wonder, 'are you mortal or a fairy?'

She looked up and smiled at him. 'I am a nixie, the spirit of this spring,' she said.

He could not take his eyes from her, thinking that nowhere in the world could there be another so beautiful as she; and for her part, she liked him well.

All that day they sat beside the stream and talked; and as the shadows lengthened and it grew near the time when he must ride away, the knight of Staufenberg felt his heart grow heavy, for he could not bear the thought of leaving the nixie. He looked deep into her lovely eyes and took her hands in his. 'Lady,' he said, 'I would have you for my wife, to live with me in my castle.'

'But I am a nixie, and no mortal maiden,' she replied. 'You must remember that.'

'I care not who or what you are, so long as I may have you for my wife.'

'I will be your wife, for I love you too,' she said. 'But there is one condition. You must be faithful and love me always, for on the day that you are faithless, you will lose both me and your life. Be untrue to me and you will die and I shall sorrow for evermore.'

He smiled. 'I would not give you one hour's sorrow, let alone an eternity of grief. And how could I fail to love you for always?' And when he spoke them, he meant the words.

So the knight of Staufenberg took the nixie to his castle and there he married her; and though she was a stranger, no one asked or guessed whence she came.

In the castle of Staufenberg they dwelt together in happiness; and, because it was a time of peace in Germany, for several years they were not parted for even a single day. But after a while, happy though he was, the knight began to think with regret of past battles he had fought and honours he had won in the service of great lords.

So when one day he heard how the king of France was seeking knights to serve him in his wars, he said to his lady, 'Though it would mean that we should be parted for perhaps a year or more, in France I could win both fame and riches. And of riches, God knows we have little enough; while of good fame no man can ever have too much. And though the parting would be hard, in the end I would return to you, more worthy of your love than before.'

And because she saw his eagerness and longing, and because she loved him with her whole being, she kissed him and said, 'Go, but do not forget me when you are in France.'

Joyfully the knight of Staufenberg rode into France at the head of his men and offered his sword to the king. For more than a year he served the king of France with such skill and courage that he rose high in his favour and won both renown and riches. But whenever the king would have sworn him to his service for all his lifetime, the knight would only say, 'My home is in Germany, lord king, and there I hope one day to return.'

The king pondered in his mind how he might keep the knight of Staufenberg about his court, for he valued him greatly; and when he found that the promise of further riches would not move him, he offered him the hand of his youngest daughter.

The knight was amazed. He did not love the princess,

though she was fair enough, but he was dazzled by the honour done him. That a king should give his daughter to a poor knight from another land, was unheard of. The knight had not forgotten his lady, far away in the castle of Staufenberg, but the temptation was too much for him, and he could not resist it unaided.

Seeking comfort and advice, he went to a priest and told him of the nixie and of the honour shown him by the king. 'Strengthen me, father, and tell me what I should do,' he begged.

The priest looked grave. 'My son, it was a grievous sin of yours, to marry with a nixie. Such water folk are lost creatures, and evil, and doomed for ever. For such a sin as yours the penance should be severe; but if you will cast all thought of this nixie from your mind and marry our good king's daughter, be her true and faithful husband as you are her father's true and faithful knight, then your sin will be forgiven you and you may live in peace with heaven for all your days.'

Although, however hard he tried, he could not think the nixie to be evil, the knight of Staufenberg accepted the judgement of the priest; partly because it forwarded his ambitions, and partly because he sincerely believed that a priest was bound to be right in such matters. He went to the king, received the hand of his daughter and promised his loyal service for all his life.

With great pomp and with the rejoicing of the people, the marriage was celebrated. But as he stood in the church beside his bride, the knight of Staufen-

berg felt his heart grow cold, and there was a great heaviness in all his limbs. And when they laid the princess's hand in his, his fingers were like ice.

As the wedding procession left the church, its way lay across a little stream where there was a narrow bridge where two could not ride abreast. The summer's day was calm and sunny and the stream no more than a trickle of water; but as the knight of Staufenberg rode on to the bridge, the water rose and washed over his horse's hoofs. He knew at once what his fate would be, but he did not turn back from it. He spurred on the terrified horse to half-way across the bridge, and there, as a great wave swelled up from the stream, he slipped from the saddle into the water to meet it; and so was drowned.

And in the very same hour, in the castle of Staufenberg, the lady of the castle was heard to give one great cry and no more; and from that moment she was never seen again.

XI

The Seven Mice

A LONG time ago, in a village on the isle of Rügen, in the Baltic Sea, there lived a widow with her seven children. Seven little girls they were, all dressed alike, from the eldest to the youngest, in gay, coloured dresses and clean, white aprons and tiny scarlet caps. Usually they were good children, well behaved and obedient; and if ever their mother had to go out and leave them alone, they would play together quietly until she returned.

But one Good Friday, as their mother set off for church, she said to them, 'Be careful to keep out of mischief until I am home again. And, above all, do not look behind the stove.'

'We will be as good as gold, mother,' they all promised; and off she went.

At first they all sat quietly while the eldest told them

a story; then they sang songs and played games together. But when they had sung all the songs that they knew and played every game twice over, they began to grow restless and fidgety, and one by one they remembered their mother's words: 'And, above all, do not look behind the stove.'

At last one of the little girls, more daring or naughtier than her sisters, got up and slipped over to the stove and peeped behind it. There she saw a bag hanging from a nail on the wall. She was so excited by the sight that she called out to the others, 'Look what there is, hanging behind the stove!' And the other six little girls all jumped up and ran to see.

One by one they peeped behind the stove, from the eldest to the youngest; and one by one they tried to guess what might be in the bag. Then one by one, the boldest first, they each put out a finger carefully and prodded the bag; but still they had no idea what might be in it. And every moment their curiosity became greater and greater until it could not be borne any longer.

And then the very boldest of them took down the bag and seven pairs of eager little hands untied the string which fastened it, and out on to the kitchen floor poured the nuts and apples that their mother had been saving as an Easter feast for them.

'Nuts!' they cried out and raced after them as they rolled across the hearth. 'Apples!' they shouted, snatching them up. And then they all sat down on the floor before the stove and cracked the nuts and munched the apples until there was not one left.

When their mother came home from church she found the seven little girls sitting quietly together, looking as good as gold. But then she saw the apple cores and the nut shells on the floor, and when she looked behind the stove she found only an empty bag.

'You bad children!' she said. 'You are like a lot of thieving little mice!' and then—very rashly, for one should be careful of what one says upon a holy day—she exclaimed, 'I wish you might turn into mice, you wicked children!' And instantly, in the kitchen, where a moment before there had been seven little girls dressed in gay, coloured dresses and white aprons and tiny scarlet caps, there were seven little mice, each one coloured brightly upon its back and white underneath, and all of them with tiny scarlet patches on their heads.

The poor mother stared at the seven mice and the seven mice stared back at her, and none of them moved for a long, long time. And there they might have gone on standing for I do not know how long, had not a neighbour knocked at the door, and having no answer to his knock, opened it and looked in. Instantly the seven mice, one after the other, ran out of the house and through the village street and away to the fields.

The mother followed them, vainly calling to them, until she had no breath left to call; but over the fields they went, never once looking back, and on into a little wood, and there, in the middle of the wood, they stopped by a little pool. At the edge of the pool they all turned round once, as if to say goodbye, then, one after the other, they all jumped into the water and disappeared from sight.

The mother, exhausted and utterly bewildered, stopped on the edge of the pool, staring at the water into which they had vanished. And there she stood for so long that she turned to stone; and there she still stands, beside the pool in the wood. But every midnight, when the world is asleep, the mice jump out of the water, one after the other, and for an hour they dance, by moonlight or in darkness, around the stone.

And so, it is said, will it always be, until a woman comes that way who has seven sons, each the same age as the seven little girls. And if she should chance to touch the stone, it will at once come to life again, while out of the pool will spring seven little girls, each one with her gay, coloured dress, her clean white apron and her tiny scarlet cap. And each little girl will take a little boy by the hand and away they will walk together, two by two, with their mothers coming after them. But when that will be, I have no idea.

Meanwhile, if you are ever on Rügen, you might look for the Mousepool with the stone beside it.

XII

Reineke Fox

In all the forest there was no other animal so heartily disliked by his fellows as Reineke Fox. They hated him, let it be said, not only because of his cunning, wicked ways, but because he was too clever for them, and everyone, man or beast, hates to be made a fool of. This is the story of what happened when the animals decided that Reineke should be brought to justice, to pay at last for his crimes.

It was springtime, and King Lion, lord of the forest, held court in the woods. To his court came all the animals and birds—all save Reineke Fox. He remained safely within the earthen walls of his fortress in the hillside with his wife, Vixen Ermelind, and his two cubs, for he had heard rumours of what the other animals intended, and thought it prudent to keep away.

When King Lion had greeted the animals, he asked if there were any there who sought the king's justice against any other; and immediately Isegrim the Wolf spoke up, 'Lord king, the crimes of Reineke Fox are many and it is time that he paid for them. I among others have much cause to complain of him, for, coming upon my cubs sleeping in the sunshine, he attacked and would have slain them all had I not come in time to save them. I say, lord king, that he should be brought before us, to answer to our charges.'

Then the Hound said, 'Lord king, last year he stole my store of bones and a fine sausage that I had laid up for the lean winter months, so that I went hungry.'

The timid Hare told how Reineke had promised to teach him his prayers, and then, when his eyes were reverently closed as he repeated the pious words, Reineke had snatched him up in his strong teeth and would have carried him off to Ermelind as a fine meal for the cubs, had not the Panther seen all that had happened and speedily come to rescue him.

And so it went on, with animal after animal speaking against Reineke and no one to say a good word for him save Grimbart the Badger. But Grimbart said with indignation, 'It is unfair to Reineke that Isegrim, his worst enemy, should be allowed to speak against him and be believed. For the words of an enemy are always prejudiced. And as for our friend the Hound, the sausage he lost was one he had stolen. What right has a thief to complain when someone else robs him?'

There is no knowing whether perhaps King Lion might not have been moved by Grimbart's words, for at that moment a sad procession came in sight, Henning the Cock, and after him two of his hens who carried between them a leafy branch on which lay the lifeless body of another hen. 'Justice, lord king,' crowed Henning, 'justice against Reineke Fox the murderer. There on that leafy bier lies Kratzefuss, my favourite wife, who laid more eggs each year than any other hen. The pride of the farmyard she was, and it was Reineke Fox who came by and killed her this very morning.'

This last crime of Reineke so angered King Lion that he gave a great roar of rage and declared, 'This villain must come before us to answer for his deeds. Braun the Bear shall go to fetch him to our court.'

At once Braun set off for Reineke's fortress in the hillside, and there he found Reineke sunning himself at the entrance. 'You are summoned forthwith,' he said, 'to the court of our king, there to answer for your crimes.'

'Alas,' said Reineke, 'who has slandered me in my absence? Oh, that I had been at court today! No one would have spoken falsely against me, had I been there. And at court I would have been had I not—it shames me to admit it, Master Braun—had I not overeaten yesterday so that this morning I had such a pain inside me that I could not for the world have dragged myself to court.' And Reineke closed his eyes and groaned most realistically.

'Overeaten?' growled Braun suspiciously, remembering Kratzefuss the Hen. 'What did you overeat?'

Reineke opened one eye and looked up at him. 'Why, good Master Braun, only simple peasant's fare, such as a fine noble beast like yourself would despise: but it is good enough for a poor fox like myself. No more than honeycomb, Master Braun, no more than honeycomb, but I thought it good and ate too much of it, greedy wretch that I am.'

'Honeycomb?' said Braun, and his little eyes glittered. 'Honeycomb! Simple fare indeed! There is nothing I love better.'

Reineke sat up. 'Why, good Braun, if I had only known! Why, I could show you, any day, a fine store of honeycomb, enough for ten or twenty bears to feast on.'

'If I might have enough of it for one bear, I would be satisfied,' said Braun.

'Why, so you shall, good Braun, if in return you will speak for me to the king when I appear before him to answer my accusers. For your word carries weight.'

'I will speak in your favour, Reineke, never fear. Only, I beg of you, show me where that honey is.'

'Come with me,' said Reineke, 'and the honey will be yours.'

And so they set off together, Reineke well pleased with his cunning and Braun eager for his feast.

On the edge of the forest stood the woodcutter's cottage, and to this cottage Reineke led Braun. Close by lay the trunk of a huge oak tree, felled by the woodcutter only the day before. He was meaning to split it all along its great length, and to this end had driven a wedge into the crack which he had already made. 'See, in there, friend Braun,' said Reineke, 'inside that crack, is the honey.'

Braun went forward eagerly, his nose twitching to catch the scent of his favourite food, and eagerly he thrust his paws deep into the crack. Immediately, Reineke knocked out the wedge and the crack closed, and Braun was caught fast, like any thief in the village stocks. In vain he appealed to Reineke to release him, promising him his friendship and his help for evermore; Reineke only leapt about him delightedly, saying, 'You see what happens to anyone who tries to take me, me, Reineke Fox, to justice.' Until at last the woodcutter, roused by the noise, opened the door of his cottage and came out to see what was the matter.

'It is time I was gone,' said Reineke. 'Enjoy your honey, friend Braun.'

When the woodcutter saw Braun caught fast, he took up his cudgel and ran for the village, calling to all the villagers to come and help him kill the bear that he had found. They hurried back with him, young and old, bringing pitchforks, sticks and cudgels, and all fell on Braun at once, so that he would surely have lost his life had he not made one last desperate effort and freed himself, leaving behind him in the tree trunk all the fine long claws from his front paws. As fast as ever he could, poor Braun ran off into the forest, the shouting villagers pursuing him as far as they dared.

When Braun limped back to King Lion's court alone, without Reineke, and told his story, loud was the animals' indignation against this latest villainy of the Fox.

'He shall be brought to justice,' roared King Lion. 'But whom shall we send to fetch him here, now that Braun, the biggest of you all, has failed?'

'Send Hinze!' they cried. 'Send Hinze the Tom-cat! He is small, but his wits will be a match for Reineke's cunning.' So they all said, while Hinze washed with his paws behind his ears and pretended not to be pleased that they praised his wits.

So Hinze went to Reineke's fortress in the hillside, walking daintily, his tail held high; and he found Reineke lying before the entrance. 'Good day to you, Reineke. You are summoned to appear before our king, to answer for your crimes.'

'Why, friend Hinze, I am glad indeed that you have been sent to fetch me, for with you I have no fear to travel. Yesterday they sent Braun to me, so rough, so blustering, that I was afraid to go with him. But I know that I can trust you. You shall be my guest for tonight and tomorrow at dawn we will set off.'

'We could travel well enough by moonlight,' said Hinze.

'At least let us sup first,' said Reineke.

Hinze considered the offer. 'What can you give me for supper?' he asked at length.

'Lovely rich honeycomb, dripping honey,' replied Reineke promptly.

'Honey! I cannot abide honey. Have you no mice in your home?'

'Alas, no,' said Reineke. 'But tell me, Hinze, are you truly so fond of mice?'

Hinze's whiskers bristled at the thought. 'My very favourite dish.'

'Then I suppose you have caught all the mice from the parson's barn?' asked Reineke.

'The parson's barn? I have never been near it. Are there many mice there?'

'Scores and scores of them, and all grown fat on the parson's corn.'

'I must visit the parson's barn.' Hinze began to purr. 'You must show me the way.'

'This very night I will show you,' promised Reineke, 'before we go to court.' And so he did, as soon as the moon was up.

'If the mice are as fat as you say,' said Hinze, as they set out, 'I might put in a good word for you to the king, Reineke.'

'I should be evermore your debtor,' said Reineke.

Now, in the parson's barn were, not mice, but the parson's hens, asleep on their perch, and often enough, in the past, through a hole in the wall, Reineke had sneaked in and stolen a fat hen. But of late the parson had set a trap beside the hole to catch the thief, and knowing this, Reineke robbed there no more.

'There is the hole,' said Reineke, when they stood outside the barn. 'Through there you may enter and catch as many mice as you will.'

'Do you take me for a fool, Reineke, that you expect me to go alone through a dark hole into an unknown barn?' Hinze was scornful.

'No, indeed, good Hinze. But I thought that you were brave, and I shall be coming after you. So why delay? Caution is a cowardly virtue, good Hinze.'

'I am neither a fool nor a coward,' said Hinze indignantly, 'and you are a fool if you take me for either.' And he slipped through the hole and was caught fast in the noose of the trap. 'Quickly, Reineke, quickly! I am caught fast in a trap.'

'Who is the fool now, Hinze? Do the mice taste good?' Reineke thrust his head through the hole, delighted by Hinze's struggles.

What with the miaowing and spitting of Hinze and the jeers of Reineke, the hens set up so great a squawking that it woke the parson and he leapt out of bed and came hurrying to his barn with a lantern in his hand.

'It is time I was gone,' said Reineke. 'Enjoy the mice, Hinze.' And he was away back to Ermelind and his cubs.

'Praised be the Lord!' said the parson. 'The thief is caught!' And he went to fetch a sack to put Hinze in, that he might throw him in the pond. And Hinze ceased his struggles and began to gnaw through the cord.

By the time the parson came back, the cord was half severed. He came close and bent to pick up Hinze and Hinze scratched him on the bare leg. The parson yelled and knocked over the lantern and clapped his hand to his plump calf. By the time he had recovered, Hinze had bitten quite through the cord and was away through the hole in the wall of the barn with the noose still around his neck and a length of cord trailing behind him. And, once again, Reineke had not been brought to answer for his crimes.

But Reineke could not escape justice for ever, and at last there came a day when he stood before King Lion and heard his fellow animals accuse him, one by one. When they had done, King Lion frowned most terribly at him and growled, 'What have you to say in your defence, most wretched villain?'

Reineke Fox looked about him and saw no single friend in sight save Grimbart the Badger, but he answered bravely enough, 'There was a time, lord king, when you were kind enough to call me friend, and were well pleased with my counsel. But now you are only ready to hear false accusations from my enemies. I am innocent, yes, innocent indeed, of all their unjust charges. And I am ready to prove my innocence in fair fight against any one of my accusers.'

The Hare kept silent and hoped no one would notice him, while Henning the Cock edged away and the Hound became very interested in some distant sound that only he had heard, Braun looked at his clawless paws and even Hinze seemed doubtful, but Isegrim the Wolf stepped forward boldly to take up the challenge. 'I say that you are a murderer and a thief and a villain, and I am ready to prove my words at any time and in any place our king may name.'

So King Lion named a time and a place for the combat and Isegrim was well pleased and all his friends were confident; but Reineke trembled and thought, 'He is far larger and fiercer and stronger than I, what hope have I of victory?' But for all his fear, he did not lose his cunning, and he set himself to think and to devise a way out of his plight. And, once again, his cunning did not fail him. On the day before the combat he shaved off all his thick red fur and oiled his body until he was as slippery as any fish, and on the day of the combat itself, he soaked his fine white-tipped brush in vinegar. 'Now let Isegrim beware,' he thought.

When he presented himself before King Lion and all the other animals who had gathered to watch the combat, loud were their jeers at the spectacle he offered, shaven and glistening with oil, and with his brush all draggled and wet; but he was not in the least abashed by their scorn.

Isegrim came forward, fierce and confident, and all the animals shouted aloud for him, for they counted on him to avenge their wrongs.

King Lion gave the word, and the combat began. With a great growl, Isegrim rushed upon Reineke, but Reineke ran in between his legs and away, before Isegrim knew where he was. But soon Isegrim was after him again, his huge teeth bared and his tongue lolling. And Reineke twisted and turned and dodged and ran from him until Isegrim began to tire, but he never seemed to catch him, for Reineke was small and quick and very slippery from the oil. And as Isegrim grew slower, so Reineke grew bolder, and every now and then he waited until the Wolf was almost upon him, when he would turn and give a flick of his brush as he ran off, scattering drops of vinegar into Isegrim's face, or he would kick up the dust into his eyes, half-blinding him.

But in time Reineke, too, grew weary from the fighting, and Isegrim, seeing this, saved all his strength for one great leap, and he hurled himself upon his little enemy and bore him to the ground. 'Now at last your day is over, villain!' he growled triumphantly.

Reineke, stretched out flat upon the ground, pinned down by the Wolf's paws, with the Wolf's great weight upon him, thought that his last hour had indeed come. 'Alas, of what avail is my cunning now?' he thought. 'I shall never see Ermelind or my cubs again.'

But Reineke never gave up hope too easily, and he began to speak fair to the Wolf. 'Dear Isegrim, give me my life and I will always serve and honour you. We should be good friends, being close kinsmen. You would find my cunning of great help to you. Have mercy on me, cousin Isegrim. Everyone admires a generous victor.'

But Isegrim only answered, 'This is one time, my Reineke, when you will not talk yourself out of trouble.' And he opened his mouth wide.

Yet Reineke, despairing though he should have been, had not yet given up all hope, and when he saw the great jaws above him coming closer, he made one last effort and caught Isegrim's long red tongue between his teeth and held on with all his remaining strength.

Isegrim howled, but Reineke did not let go. Isegrim tried to break away, and Reineke slipped out from underneath him, still holding fast to the long red tongue. 'Yield, Isegrim, and admit the accusations false, or I bite it off.'

So Isegrim yielded, and King Lion declared that Reineke was the victor and thereby proved innocent of all the charges brought against him; to the great indignation of all the other animals, all save Grimbart the Badger.

And King Lion considered the matter gravely, and thought how Reineke must be not only innocent but also greatly virtuous, that he, so small and weak, could have prevailed in combat against so large and strong a foe. And King Lion reflected further and decided that Reineke must surely be wise as well as virtuous, and he once more called him his friend and, indeed, took him for his chief counsellor and always listened favourably to his advice, ignoring the words of far better animals; so that Reineke and Ermelind and their two cubs all prospered exceedingly.

And so it may be seen, from the story of Reineke Fox, how it is not always virtue which is rewarded, nor the good who are always honoured. Yet, for all that, since the beginning of the world, there have been many less likeable villains than Reineke Fox.

XIII

Eppelin of Gailingen

IT is told how, in the middle ages, there lived at Gailingen, near Schaffhausen, on the Rhine, a robber knight named Eppelin. This Eppelin was wild and bad, hard to his friends and ruthless to his enemies. Indeed, he had few enough friends and counted most men his enemies, and he loved little in heaven or on earth save his horse. His horse was the finest for many miles around, and there was a deep trust and understanding between it and its master, so that it was almost as though the horse knew what Eppelin wished of it and would seem to obey him even before he had given his commands. And for this reason, it was a most valued companion and helpmate to Eppelin on his raids.

Eppelin of Gailingen would lie in wait, well hidden by the trees and undergrowth, beside the road to

Nuremberg, where the merchants passed with their rich wares for the great fair laden on their strings of packhorses, and along which they returned home again with their horses travelling lighter but their money bags well filled. When a cavalcade appeared in sight that seemed ill protected and worth the pains, Eppelin and his men would ride out from the bushes upon it, swords drawn and shouting horribly; and in a few minutes they would have plundered it of all its money bags or of the finest of its merchandise, and be away again almost before the unfortunate merchants and their servants could strike a blow in their own defence.

This went on for years, until Eppelin's became the most hated and most feared name in all the countryside. Many were the men the lord of Nuremberg sent out to take the robber knight, but always they returned to his castle empty-handed, while Eppelin grew daily richer and more insolent.

But no one can have good fortune for all the time, and at last there came a day when Eppelin of Gailingen was captured and taken bound to Nuremberg. In the grim castle of Nuremberg, with its five-cornered tower, stripped of his armour and his sword, his helmet and his spurs, Eppelin stood in chains before the lord. 'You have troubled the countryside long enough and made life a danger to honest folk,' said the lord of Nuremberg. 'Now you shall trouble it no longer.' And Eppelin, condemned to be hanged, was flung into the deepest dungeon of the castle; while his horse was led to the stables, a welcome prize for the lord of Nuremberg.

A day or two of captivity, while all was prepared for the hanging and people flocked to the castle from the villages around, passed all too quickly for Eppelin in his dungeon; and one fine morning he was led out to die, unrepentant and determined to be defiant to the end.

In the middle of the courtyard stood the gallows that had been built for him, with the hangman and his assistants standing by; the lord of Nuremberg was there with all his court; there, too, were many townsfolk and villagers, all come to see the hated robber die.

Eppelin looked about him scornfully and gave a laugh. 'All these people,' he jeered, 'so many folk all gathered here for my sake: and not a single friend.' He laughed again, harshly.

But he was wrong. From the stables his horse had heard his voice and whinnied in pleasure, thinking that the master whose company it had missed for two long days, was come at last. And on Eppelin's lips the mocking smile faded, as he remembered that in all Nuremberg he had still one friend; and he thought how it would be a sorry thing to die without bidding farewell to a loved and trusted comrade.

So he humbled himself and knelt before the lord of Nuremberg and asked a boon before the rope was put about his neck. 'Grant, lord,' he said, 'that before I die, I may ride my horse yet once again. Just to ride once or twice around the courtyard, that is all I ask.'

The lord of Nuremberg looked about him at all his men at arms and at the eager crowd, and beyond them to the high, broad walls and the barred and guarded gates, and he could see no way of escape for Eppelin. And, too, he thought how he was not disinclined to see put through its paces the fine horse that was now his own. So he granted the request.

The horse was brought from the stables and saddled and bridled. Eppelin had his chains struck off and was allowed to mount. Slowly he walked his horse once around the courtyard, curbing its eager pace, while the crowd watched and murmured its admiration of a horse and rider so well matched. The second time around the courtyard, Eppelin allowed the horse to canter, feeling once again an old joy coursing through him as he rode; while the lord of Nuremberg leant forward to watch and the murmur of admiration grew louder. The third time around the courtyard, Eppelin gave the horse its head, and, fresh from the stable and its days of waiting, it galloped, sparks flying from its hoofs against the cobbles, and Eppelin said to himself, 'If only this might go on for ever.'

Then suddenly he thought fiercely, 'It would be better to die as I have lived—in my own way—than at the will of another. Better to say how and when for myself, than to await the pleasure of any lord. Far better to die quickly with my head cracked open on the cobblestones, than to die slowly on the end of a hangman's rope. And far, far better that my horse should lie dead with a broken neck than belong to any other man.' And he reined in the horse and turned it to face across the centre of the courtyard, and he bent forward and spoke to it. Like an arrow from a bow, the horse responded, right across the courtyard, past the gallows in the middle and on, straight towards the wall opposite, too fast for anyone there to realize what was happening.

Only a few yards from the wall, Eppelin spoke one word to the horse, and it leapt straight at the stone barrier towering above it, while the people standing there at the foot of the wall cried out and scattered as the hoofs flashed over them. Eppelin closed his eyes and set his teeth and waited for the sharp pain and the blackness that would follow, and felt instead the cold freshness of the wind in his hair and heard the cries of dismay from below him; and he opened his eyes to find that the impossible had happened and the horse was standing, all four hoofs safely on top of the broad wall.

Eppelin did not waste a second to look behind him. Another word, and the horse leapt fearlessly out, over the wide moat below, reaching the ground no more than a few inches beyond the deep ditch.

Long, long before the amazed guards had opened the gates of the castle and the men at arms had ridden forth in pursuit, Eppelin of Gailingen, the robber, and his brave horse were far along the road from Nuremberg.

XIV

Big Hermel

A FEW score peasants once lived in a village. They were poor and had little to eat and no fine clothes to wear, and most of them worked hard from dawn to dusk; but their overlord, if not exactly kindly, was at least just; and though they may have grumbled amongst themselves now and then, as people will, on the whole they were happy enough.

But one day their overlord died and another took his place. The new lord was harsh and cruel and demanded far more than his rights. And if they had worked hard before, they were now forced to work harder; and where they had had little to eat, they now had less; and now they had not even their rough homespun clothes to wear, but they went in rags and shivered pitiably when the cold winds blew.

There was in that village a simple young fellow named Big Hermel. He was twenty years old and six yards tall, and strong and broad-shouldered into the bargain. A good man to have as a friend and a bad man to have as an enemy, one might think: but it was not so. For though Big Hermel was good-natured to a fault and always ready to laugh and joke with everyone, he would have made a poor sort of friend for any man, for, though he never did any harm to anyone, nor would he ever stir a finger to help anyone, either. And he had never done a day's work in all his twenty years. Indeed, for laziness, there probably never had been Big Hermel's equal.

This had not mattered very much while the old lord lived, but now, with the new lord's bailiff harrying and nagging the whole time, and his men always ready to grab their last measure of wheat or exact their last hour of work from the unfortunate peasants, such laziness was more than the villagers could put up with.

When it came to harvest time, and everyone, man, woman and child, went out to the fields to get in the lord's plentiful harvest before they could go to their own miserable little fields and cut their own sparse corn, Big Hermel just lay in the grass on his back, staring at the blue sky or sucking a straw.

The villagers could stand it no longer and they turned on him. 'We have put up with you for long enough, Hermel,' they said. 'You never do a stroke of work for your keep, and, to make matters worse, you eat as much as four men. Either come and help us with the threshing now, or you can go without this winter.'

Big Hermel sat up and stretched himself. 'You want me to help with the threshing, do you? Why, that's easy. You should have told me before.' He got up and strode off to the wood. There he tore up by the roots two saplings, an oak and a pine, and bound them together to make a threshing flail. With this flail he threshed all the grain in no more than one half-hour.

The peasants were delighted to have their work done for them; but the bailiff and the lord's men were none too pleased. 'How can we be sure,' they said amongst themselves, 'that this Hermel will not think to use his great strength against us, to resist our demands? There are few enough of us, and I, for one, have no fancy to be threshed like corn.' They thought about the matter for quite a time, and then they said, 'We cannot be sure of what this Hermel will or will not do. All we can be sure of, is that we must get rid of him, and soon.'

So the bailiff sent for Big Hermel and said to him, 'You are a fine strong fellow, and I have the very task for you. The river has been dammed at a certain place by a fall of rock. I want a fine strong man like yourself to go and clear the stream.'

'That will be easy,' laughed Big Hermel, and off he went to do it.

But the bailiff followed him with the lord's men and they stood on the bank of the river to watch. And when Big Hermel was in mid stream and up to his shoulders in water, they all picked up large stones and flung them at him, hoping to drown him in the river.

But Big Hermel only brushed the stones aside, looked up at the sky, looked across at the bailiff and the lord's men, grinned cheerily at them and said, 'Those were the biggest hailstones I have ever seen.'

So the bailiff sent for a huge millstone, and he and the lord's men took it up and flung it at Big Hermel. But he caught it as it came and put it round his neck with his head through the hole in the middle, laughed and waved to the bailiff and the lord's men on the bank, saying, 'Thank you for the fine new collar!'

Then the bailiff sent for a great iron bell, as much as three men could lift, and he and the lord's men heaved it up and flung it at Big Hermel. Big Hermel saw it coming and moved so that it fell right upon his head. He set it straight, turned to the bailiff and the lord's men, patted the bell and shouted to them, 'You have given me a fine new hat. Many thanks!'

Then the bailiff and the lord's men looked at one another, and went back to the castle and told their lord about Big Hermel. 'Unless we are rid of him, lord,' they said, 'he will destroy us all.'

'I will find a way to be rid of him, never fear,' said the lord, and he set himself to think.

The next morning he sent for Big Hermel. 'Hermel,' he said, smiling in as friendly a fashion as he could manage, 'I am rather short of money at present. I want you to go down to hell and ask the Devil to give you a sack of gold, as much as you can carry away with you. Bring it back here to me and I will give you a share of it. We shall have a fine time spending it, I have no doubt. You are the best man to send to the Devil, since you can carry far more than anyone else.'

So Big Hermel found the path which led down to hell, and he went along it for more than an hour, until he came to a great iron door. The door was locked, so he knocked on it. However, no one came to open the door to him, so he became impatient and knocked again, more loudly. But he knocked so hard that the whole door fell in with a mighty clatter, and Big Hermel stepped over it and into a wide, high hall, well lighted by a hundred flaming fires or more.

The din brought the Devil running. 'What do you want?' he asked, very angry at the disturbance.

'A sack of gold for my overlord, as much as I can carry away,' said Big Hermel.

'What impertinence!' stormed the Devil, his tail switching furiously. And he took Big Hermel by the neck to throw him out of hell. But with one finger Big Hermel gave the Devil such a blow that his teeth rattled in his ugly head.

The Devil let go of Big Hermel quickly and did some fast thinking instead. 'See here, my friend,' he said, after a moment. 'If you can win three wagers against me, I will give you the sack of gold you ask for. But if you lose, your soul is mine. Agreed?'

'Agreed,' said Big Hermel.

The Devil took up a huge hunting horn and blew on it a blast so great that the sound of it put out seven of the fires of hell. 'Now better that,' he said, handing the horn to Big Hermel.

Big Hermel put the horn to his mouth and blew. And Big Hermel's blast on the horn put out all the other fires of hell, up to a hundred or more.

The Devil raised his eyebrows. Then, with both his hands, he took a stone as large as a bakehouse and flung it twenty yards up. 'Better that if you can,' he said.

Big Hermel took the stone and tossed it from one hand to the other as though it had been a feather-filled ball. Then he peered up into the rafters above him. He put the stone down again and said to the Devil, 'If you can wait a minute or two, I will just run up to the earth again and pull up two oak trees to shore up the roof of hell. Otherwise, when I throw up the stone I shall probably break the arches and the roof will fall in on us. And there would be an end of us both, which would be a pity.'

He turned to go, but the Devil, looking rather pale, stopped him. 'No! Wait! You win the wagers, Hermel. I will give you your sack of gold.'

So the Devil gave Big Hermel a huge sack full of gold, and Big Hermel, very pleased with himself, strode away up the path from hell, whistling merrily.

When he arrived at the castle with the gold, the lord and his bailiff and all his men could hardly believe their eyes. They hid their disappointment as well as they might, and the lord said to Big Hermel, 'It is a good day's work you have done for me today. Tomorrow you can take home your share of the gold; but you must stay and feast with us tonight.'

'Many thanks, lord,' said Big Hermel. 'I shall enjoy that.'

Up at the castle there was more food and drink than Big Hermel had ever seen in his life before, and as much of it as he wanted was his. He ate enough for seven and, thanks to the lord's men, who kept on refilling his cup, he drank enough for fourteen. And when he could not hold another drop more, he fell off the bench, slid under the table and began to snore.

'This,' said the lord, 'is our last chance. And this time there must be no mistake.'

They carried Big Hermel, still snoring happily, out to the courtyard and laid him down. Then they piled brushwood and logs high around him, and when they thought there was enough, they set fire to the heap. The wood burned and crackled merrily, and the flames rose higher and higher, but Big Hermel went on snoring and never noticed a thing.

The lord and his bailiff and his men stood around watching and rubbing their hands together with glee. 'This time,' they chuckled, 'we are really rid of him.'

But the great iron bell which Big Hermel wore on his head began to grow hotter and hotter, and at last, just as his hair was beginning to singe, the heat of the bell awoke him. He saw the flames all about him and sat up. 'Why, what is this?' he exclaimed. He stood up amid the fire and saw, beyond the flames, the lord and his bailiff and his men, all laughing fit to split their sides.

It took him a moment or two to piece things together, but when he had done so, for the first time in his life, he was angry. 'What a mean trick to play on a man who has had too much to drink,' he shouted. 'If that is the way you treat your guests, then the world would be a better place without you all.'

He gave a great leap and leapt right out of the fire, tore up by the roots a tree which grew in the courtyard, and then, as though he had been threshing corn, he laid about him with the tree, just as he had done at the harvesting. And that was the end of the lord, his bailiff and all his men.

'Well, so that is that,' said Big Hermel, looking

around. 'Now perhaps I will be left in peace, since there is no one to make me do any more work.'

And he shouldered his tree, took up the sack of gold and strode out of the castle and back to the village.

After that the village had a new lord. He was a kindly, well-meaning soul, and his men took after him; while his bailiff was as easy-going as a bailiff can possibly be. Which was, perhaps, as well for them.